Clearing Vietnam

Anatomy of a U.S. Army Land Clearing Team

A Memoir

By Terry T. Brown
Sp-5, United States Army

A detailed Chronicle of events, trials, and hardships during 1968 and 1969, in an 18-month Vietnam tour of duty.

20th Engr. Brigade   86th Engr. Bn.   62nd Engr. Bn.

All rights reserved. No part of this book shall be reproduced or transmitted in any form or by any means, electronic, mechanical, magnetic, photographic including photocopying, recording or by any information storage and retrieval system, without prior written permission of the publisher. No patent liability is assumed with respect to the use of the information contained herein. Although every precaution has been taken in the preparation of this book, the publisher and author assume no responsibility for errors or omissions. Neither is any liability assumed for damages resulting from the use of the information contained herein.

Copyright © 2007 by Terry T. Brown

ISBN 978-0-7414-3974-1

Printed in the United States of America

Cover artwork, 'Engineers Clear The Way', by Dale Gallon, appears through the courtesy of Gallon Historical Art, Gettysburg, PA www.gallon.com

INFINITY PUBLISHING
1094 New DeHaven Street, Suite 100
West Conshohocken, PA 19428-2713
Toll-free (877) BUY BOOK
Local Phone (610) 941-9999
Fax (610) 941-9959
Info@buybooksontheweb.com
www.buybooksontheweb.com

Contents

	Prologue	1
Chapter 1	Starting Out	3
Chapter 2	Bearcat	8
Chapter 3	Land Clearing	30
Chapter 4	Security	48
Chapter 5	Field Maintenance	76
Chapter 6	Monsoon	90
Chapter 7	Chow	97
Chapter 8	Misfortunes of War	111
Chapter 9	A Veritable Zoo	123
Chapter 10	Stand Downs and R&R	141
Chapter 11	In Convoy	157
Chapter 12	Exiled	188
Chapter 13	Shepard's Way	213
Chapter 14	Up in Smoke	227
Chapter 15	Redemption	239
Chapter 16	Flight of the Freedom Bird	263
	Epilogue	280

Acknowledgments

Many thanks are extended to all those who, in some way, contributed to the overall composition of this memoir. Without their added input and support, its content would surely have been less detailed and informative than its present form. For technical backup, much appreciation goes out to my former unit's Chief Warrant Officer, George Ingledew, who helped to verify and point out some of the mechanically related facts and details that I had long forgotten, regarding our heavy equipment. Additional gratitude goes to my Platoon sergeant, Paul Spitdowski for helping me get some of the incidental facts straight; and to his wife, Janet, who kept prodding me to continue writing and ultimately finish what I had curiously started. Also, to (Capt.) Jim Nixon, for passing along a photo of the historical artwork image that now graces the cover of this book (*courtesy of Gallon Historical Art in Gettysburg, PA*). Additionally, to my friend, Ronald Kreps, for his technical assistance with some of the various photographs and other images that were added to this book. And lastly, thanks to Google, for being my research vehicle, in retrieving key information that helped to tie everything together.

Dedication

This memoir is dedicated to all the courageous and determined men associated with Land Clearing operations in Vietnam, with special recognition toward those who suffered the ultimate sacrifice, by losing their lives in the process. These men were an elite group of combat engineers, who always found a way to get the job done in spite of adverse and abnormal conditions. Their shared glory is in knowing that the lives of hundreds, if not thousands of other military personnel, along with numerous Vietnamese civilians, were spared as a direct result of their combined efforts in removing heavy jungle growth from along key roadways and from within various area strongholds, while effectively denying the enemy uncontested access toward carrying out their signature hit-and-run ambushes.

Additionally, this compilation of memories is gratefully dedicated to one of my Company Commanders, Captain John J. "Jack" MacNeil, who, through his efforts to lead us into hazardous enemy infested jungles, took extra precautionary measures to help ensure our safekeeping, thereby lessening the amount of killed or wounded among us.

I saluted a nobody.
I saw him in a looking glass.
He smiled-so did I.
He crumpled the skin on his forehead,
frowning-so did I.
Everything I did he did.
I said, "Hello, I know you."
And I was a liar to say so.
Ah, this looking-glass man!
Liar, fool, dreamer, play-actor,
Soldier, dusty drinker of dust-
Ah! He will go with me
down the dark stairway
when nobody else is looking,
when everybody else is gone.
He locks his elbow in mine;
I lose all-but not him.

- Carl Sandburg

Prologue

This detailed written account, as recalled from a period during the height of the Vietnam War, is a collection of personal memories and observations, along with other related subject matter. It is a revealing compilation of extraordinary experiences, practices and encounters that occurred while serving much of my eighteen-month Vietnam tour of duty, during most of 1968 and 1969. Being almost entirely drawn from memory, after 38 years of dormancy, it is intended to show a different view of a soldier's life there, as it unfolded in my case, in describing and detailing the unusual type of work that was taken on in support of others, and in recounting some of the more finite details of conditions and incidental situations. At times, it may seem that I described some things a bit too much in depth, but in the effort of trying to place the reader there with me, some elaborate descriptions were deemed appropriate. In the interest of accuracy, special care was taken in staying as close to the facts as I remembered them, while an impartial, but critical eye helped to deter any possible notion of exaggeration or embellishment. Although my experiences there were somewhat different than those of some others, they were largely collective in many respects, as the other men around me lived through variations of the same day to day involvements.

Given all of the articles that I had later read about my former unit, as well as its sister elements in Vietnam, they had largely come to be recognized as textbook type recordings of the various logged events, along with related statistics and such, while being mainly scribed in military style writ. But, in writing this compilation of memories, I thought to insert my own experiences, as well as those of the unit, to help personalize it somewhat in providing the reader with more of an outward portal to that time period.

As a former U.S. Army Combat Engineer, I remain moved by my time spent in South Vietnam. It served to rapidly change my early innocent and naïve boyhood nature, leaving me more whole and wiser, and more engaged toward a fuller likeness of a man. Most of us just call it 'growing up,' but living, working, and surviving within a war zone causes one to do it at a more accelerated pace, and with more profound purpose. Moreover, this memoir serves to honor the skills, ingenuity, and overall tenacity of a unique combat engineer unit, whose daily arduous activities made a marked difference in changing the wartime landscape of South Vietnam, while keeping troops and civilians more secure from potential enemy ambushes.

~ One ~

Starting Out

It was mid-February, 1968 when I traveled early one morning, making the 18 to 20 mile road trip from the east bay home of my parents, out through the Caldecott Tunnel and down to the Oakland Army Base, where thousands of Army personnel, throughout each year of the war, were being processed for entry into the southeast Asia theater... namely, Vietnam.

I was just 19 and only a few months removed from high school and the warm California summer sun. Basic training and AIT were happily behind me, and now my 30-day leave was up. It was my time to report, according to my orders, and my folks were good enough to deliver me there and see me off. Within hours, I found myself departing Oakland aboard a World Airways Boeing 707 filled with anxious others, destined for a year's duty in a war torn far off land.

After re-fueling on the island of Okinawa, with 22 hours elapsed time, we finally touched down at Bien Hoa Air Base and were all escorted from the tarmac, and bussed to a holding company in the area, where I spent the next couple of days waiting for someone to claim me. We would assemble in formations twice a day to hear our names called out for whichever designated unit. Those whose names were called would soon hook up with a waiting vehicle for the road trip over to wherever their newly assigned outfit's location happened to be.

Otherwise, until claimed, there wasn't much of anything to do, except wait.

[Prior to joining the Army, oddly enough, I had never ventured outside of California *(except for a brief summer visit to the Nevada side of Lake Tahoe)*, and suddenly, within a short period of time, I found myself taking basic training in Washington State, and advanced training in southern Missouri. Upon arriving in Vietnam, I realized that I was now thousands of miles away from home; clear on the other side of the earth; and it might as well have been Jupiter, as I had quickly become isolated from the only world that I had known. Initially, I felt as if sentenced to serve my time in this strange new environment that seemed like an entirely different planet to me. Granted, as it was, I enlisted, but only to avoid the consequences of the draft. Entering the military then was the very last thing I wanted to do, short of incarceration. But, the stark reality of national conscription soon changed my tune, as my non-compliance surely would have involved a jail term.

At that time, in August of 1967, the draft was in full swing with the looming specter of the lottery then largely coming into play, while many of those my age were being randomly selected and called up to serve. Through this arbitrary selection process of picking assigned numbers from a national registry of Selective Service members, the lottery's downside revealed that one's fate was to be entirely determined by the military, once they got their hands on you via the draft. My biggest fear at the time was to get drafted and become a 'grunt' (*an infantryman*), as the news media was already showcasing the plight of some ground troops in Vietnam, with growing concerns over increased land mine casualties being highlighted regularly on the nightly TV news. At that point, as I realized the high probability of my number being called, I figured, either way, they probably already had me snared in their trap, with my fate then likely to rest more in the hands of uncertainty than anything else. About the only thing I could possibly do to avoid that unsettling plight was to try to improve my lot in some way by effectively circumventing the national lottery. In that regard, it was painfully simple, as I still had an option available to me that offered a way around the uncertainty of the draft. Enlisting in the military would actually bring about some choices that could enhance my chances of survival, while on the other hand, simply waiting around for my number to be called pretty much weakened those chances, given that I was likely to be heading over to Vietnam like so many others were at that time.

So, after talking things over with my father, he and I went to see the local recruiter, who suggested that I take a three-day written examination at the Oakland Induction Center. With the results in, the recruiter stated that I scored very high in the mechanical aspects of the test and suggested that I could easily qualify for further training in the Army Engineers if I wanted to go that route. As my father had been a 20-year Navy man, seeing WWII action in the Pacific, he related to me that this was quite similar to the Navy's Seabees and could offer some valuable work experience for the future. So, it was then that I decided to go to the Army's Heavy Equipment School, where I wound up training on Bulldozers and heavy construction Scrapers, while believing that if sent to Vietnam I'd at least have some substantial protection from those dreaded land mines.

Unlike some others who may have gone into the service for reasons of patriotism and a willingness to help fight for the cause, I didn't really care to be a part of any of it until my father actually pushed me in that direction and persuaded me that military service would be an educating and memorable life experience. Of course, he was right in a strangely urgent sort of way. So, on September 10th, while still somewhat reluctant to act, I signed on the dotted line and, after taking my physical in Oakland, took the oath of allegiance to support and defend the constitution of these United States, after which I was quickly sent on my way to basic training at Fort Lewis before fully realizing what had actually hit me.

Born in 1949, I hailed from a combined family of seven children. My step-sister, along with two older step-brothers, lost their father at an early age to a fatal heart attack. My father then entered the picture, siring four more to my mother after exiting the Navy and taking up residence on a five acre ranch, in the remote little east bay town of Alamo. We raised sheep, pigs, chickens, horses, and had a couple of milk cows for our own consumption of homemade dairy products. We also pridefully cultivated a yearly victory garden for fresh and preserved produce. While my mother was a 'devout Catholic' and third generation California native, mainly of English ancestry, my father, being of Irish and Scottish descent, came to California via the Navy and the Japanese war in the pacific, originally out of Brockton, Massachusetts. As a Lieutenant (JG.) in the Navy, he endured the torpedo sinking of his warship, U.S.S. Northampton, near Guadalcanal in 1942, having to swim a great distance to safety where a Navy destroyer ultimately rescued him along with many others, while still other men trapped below deck, went down with the ship. During the mid-50s when I was just a boy of 7, after my father had served several years as a police patrolman for the county sheriff's department, he decided to run for public office and, after a lengthy campaign, he was elected Sheriff of Contra Costa County.

My eldest step-brother, being eight years older than I, had entered the Army in the early 60s and, after later attending Officer's Candidate School with further training as a helicopter pilot, he was sent to Vietnam. He served in the Saigon area until he took a rifle round up through the bottom of his Huey gunship that entered his back and exited out through his right armpit, taking most of his bicep muscle with it, while leaving him to nearly bleed to death. His co-pilot then assumed control of the chopper and made it to the nearest field medical hospital in the nick of time. Despite the tremendous loss of blood, surgeons managed to save his life, but his right arm would never be the same again. Although it was spared from amputation, he could no longer use it for much of anything. That was just about two years or so prior to my entrance into the Vietnam theater.]

~ Two ~

Bearcat

On my second full day in country, my name was finally called and I found myself assigned to the 86th Engineer Battalion, out of a place called Bearcat, a somewhat small post located out along QL-15 highway, east of Saigon and south of Bien Hoa within the III Corp Tactical zone.

For tactical deployment and divisional troop responsibility, the land mass of South Vietnam was divided into four zonal areas that reflected strongly on terrain. I Corp was situated in the Northern Highlands, establishing the dividing line between the two battling countries at the de-militarized zone, while IV Corp encompassed the entire river delta region to the extreme south. Located between those northern and southern zonal areas, II Corp was situated in the Central Highlands area, while III Corp lay just north of the delta region, within the Central Lowlands.

At some point, a little later in the day, a clerk from the 86th showed up, driving a ¾ ton Willy's pickup truck with an O.D. (*olive drab*) green canvas covered load bed. After acknowledging his presence, I stowed my duffel bag in the back and hopped up front with the driver for the relatively short road trip over to Bearcat. About a half hour into the ride from Bien Hoa, the truck pulled off the main highway (*QL-15*) and turned onto a gravel road. I casually noticed an entry sign posted along the way, which featured an image of 'Sugar Bear' (*an animated cartoon character from an old TV cereal commercial*), with the words, *Welcome to Bearcat* written on it.

As much as I recall, Bearcat was roughly 20 miles southeast of Saigon, and about 8 to 10 miles due south of Long Binh Post, within the southern reaches of Bien Hoa Province.

In addition to being the home of the 86th Engineer Battalion, Bearcat was also home base for part of the 9th Infantry Division at that time. It also housed a few Army helicopter units, an MP Company, and a Communications Company, while the other half of the base was occupied by a Royal Thai Army Regiment, called the 'Queen's Cobras.'

The 86th was a Combat Engineer support battalion, which was attached to the 9th Infantry Division. At that time, part of the 9th Division was still in Bearcat, but, several months later, they would move down to the IV Corp area, into the Mekong River Delta locale of Dong Tam, with the 86th soon to follow them.

Arriving at battalion headquarters, I was greeted by Headquarters Company's First Sergeant and informed that, because of my particular MOS (62E20) {Crawler Tractor Operator}, I was to be assigned to a somewhat new assemblage, called LCT (Land Clearing Team), which was in the infant stages of becoming a company, after spinning off from a platoon within A Company. At the time of my arrival, this new assemblage of men and equipment was still in the process of evolving from its earlier platoon status to eventually become its own separate detachment within the battalion, just short of achieving company status. Although I had anticipated operating a standard bulldozer within a construction battalion, I found myself involved in a completely different undertaking. It amounted to dirty, dangerous demolition type work, where I operated a Caterpillar D7-E tractor that was specially equipped for cutting and clearing dense jungle growth. In applying my basic skills to this unusual and unconventional activity, I later found that it offered an altogether different type of challenge that was somewhat foreign to my earlier pre-conceived notions about heavy equipment operation.

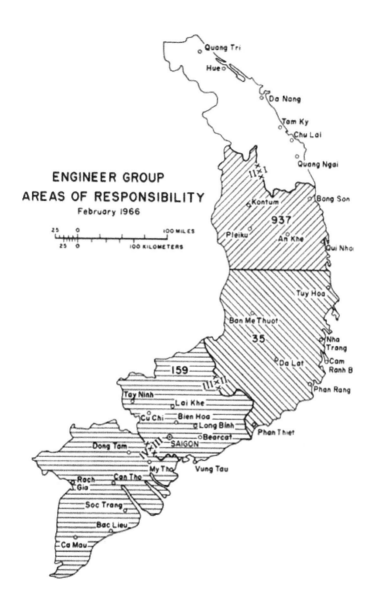

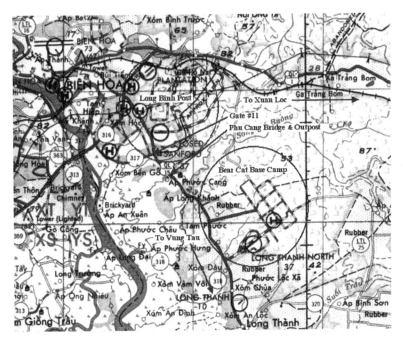

Map of QL-15 highway, with Long Binh Post and Bearcat noted, as it runs north to meet QL-1, near Bien Hoa.

The first night there, while in Headquarters Company as the turbulent days of the Tet Offensive were well underway, rockets and mortars rained in on us, and "all hell broke loose." I found myself taking cover beneath a table down on the mess hall floor alongside the mess sergeant there, before we finally found our way out to a nearby bunker, as I had KP (*Kitchen Patrol*) duty that day. The sporadic barrages lasted throughout that night and, to a lesser degree, continued for several nights thereafter. Being newly in country, I simply assumed that this must be the usual level of enemy activity for this area. But, fortunately, I was mistaken. The Tet offensive had soon after been effectively quashed and things noticeably calmed down around there, while restoring our surroundings to a more controlled level of action, with only a few occasional enemy flare-ups occurring from time to time.

After processing into the battalion, I was assigned to LCT's 2nd platoon and soon found my way over to their sand-bagged, single-story Quonset hut type barracks, which was located within the confines of A Company. I met a few of the other men there, and finally settled in to accept my new post, still not really knowing exactly what might lie ahead.

While getting a little more acquainted with my new surroundings, I learned, among other things, that privacy was more of a stateside luxury. Our latrines there were presented to us in the form of elongated outhouses that contained several toilet seats (*about 8 of them*) mounted upon a plywood-constructed shelf, and were all lined up right next to each other without even a divider between them. At the risk of sounding a bit crude, each morning seemed almost like an elimination tournament of sorts with the place being generally packed. As its foul-smelling inner atmosphere mixed with generous amounts of cigarette smoke, most of us simply resigned ourselves to the combined offending odors, while naturally focusing more on the necessity of the moment. Showering with a bunch of naked men was one thing, but sharing one's 'morning constitutional' in this way was entirely quite another. However, this kind of intimacy we all had to get used to.

Positioned below each toilet seat there, about five or six feet down, were circular metal receptacle pans that had been cut from the tops and bottoms of 55-gallon fuel drums. Each day, some unlucky soul had the unenviable task of raising the access doors behind the outhouse and pulling out all the trays of accumulated matter with a hooked metal bar that had been fashioned out of re-bar. With the lot of them all lined up behind there, as if ready for inspection, diesel fuel was generously splashed over the contents of each, and they were all set ablaze. This was a common practice that most everyone in country had utilized, toward effective elimination of solid waste, as there were no sewer systems around and septic systems were rarely seen. Maintaining a high level of sanitation was the important thing there, and burning it all off seemed to be the best way of achieving that end result, given the overall lack of more modern methods. However, the black smoke that billowed from the 'burn off' often affected our own air quality around the company area, as some of us experienced some mild nauseating effects from time to time.

As for urinals, they were also fashioned from 55-gallon metal drums and were buried vertically in the ground with about 4 inches protruding. They were punctured and laid in with gravel surrounding each one (*for drainage*), and had fine steel mesh screening applied and spot-welded to the upside 'business' end. Small, corrugated 4-foot high metal partitions were erected in front of each one for some small measure of privacy, as these urinal stations were not properly housed like the 'shitter' was; and although they were chemically treated, these urinals were located more out in the open, in a few different outskirts of the company area, where any possible residual smell of urine was mainly kept isolated from our work areas, and from the general vicinity of our living quarters.

Our movie theater in Bearcat was simply a large wood-framed, standard-sized theater screen that had been erected within the central open space, in a wide part of Headquarters Company's general area, and was shared with the rest of the 86th battalion. We regularly gathered at night to watch our 'flicks' there, out in the open air beneath the stars. Those who had folding chairs could sit out in the open, on the sandy, beach-like surface, *(the area was alternately used for volleyball games)* for a more direct view of the big screen. The 'chair-less' among us simply sat atop the hooches and sandbagged walls of the areas' perimeter, with a fair supply of canned beer and cigarettes on hand to keep us company. Our featured films were generally hit or miss, with a somewhat enjoyable one occasionally making it to the screen, along with some rather off-beat ones. I always liked the Looney Tune or Roadrunner cartoon that normally preceded it. But, most of us usually showed up for the main 'feature' anyway, and stayed at least until it became more evident that it was effectively curing our insomnia.

The nearby Enlisted Men's Club (*EM Club*) was a small, dark little bar with a few small square tables and accompanying straight-back chairs comprising the usual basic layout, and was always good for a few beers with the guys and a bowl of nuts. The movies just outside were always free, but the beer at the club was a common expense, and sometimes unaffordable as funds usually got a little lean near the end of the month with payday lying just around the corner. Given my age of 19, I was quite tickled that I could actually drink alcoholic beverages on military installations, as the Army never set a drinking age, like my home state of California's established age of 21. I guess they just plainly figured that if you're old enough to be a part of Uncle Sam's Army, then you're old enough to drink.

Outgrowing our temporary digs within A Company, the evolutionary process continued for LCT and we soon gained our own separate area, which was located somewhat closer to our motorpool out on the far edge of the battalion's fairly large, assigned parcel of dusty barren land. While a few of our out-buildings and living quarters were still under construction at the time, there was still adequate room for the bulk of us to move in and set up within our newly annexed area, where our daily business was to focus almost entirely on land-clearing operations and maintenance in and around those more troublesome areas of III Corp, where we were most desperately needed.

At the same time, the other companies within the battalion tended to focus more on their construction-related projects that were allocated in full support of the 9th division, as the 86th Engineer Battalion, in having been assigned to serving the construction needs of the

9th, had been officially designated as a 'Combat Construction' battalion, which was their recognized, primary job description. Our evolving Land Clearing detachment, although a necessary growing force in the fight against enemy insurgency, didn't always relate well with the battalion's interests towards supporting the 9th Division. We were of a different vein entirely, with land clearing being our only activity in view. Having cleared many large areas of jungle for the 9th over a period of a year or so, while conveniently operating within their main engineer support battalion, we later found that they no longer needed our help in their area of operations. The other Infantry Divisions in the various outlying areas of III Corp then began to call on our jungle cutting services in helping with their own particular vital security needs. We had often been a project in logistics for the 86th, moving our rather large convoys with all the necessary vehicles that it took to facilitate those mobilizing efforts whenever transporting our heavy equipment and other essentials from place to place in the field. But, the 86th Battalion was more dedicated to the needs of the 9th Division as they were specifically assigned to that particular duty, while still supporting the efforts and the growing needs of their company-size Land Clearing Team.

While getting more acclimated there, I had also noticed that some local Vietnamese civilians were present on post. Men served as general clean-up workers, and able bodied women and girls served as cleaning women and laundry girls for our new canvas-covered huts (*called, 'Hooches'*), and also for the wood framed 2-story barracks that some of us then occupied.

Our wood-framed, canvas covered 'hooches' are seen here in various states of completion, with a bunker established on the extreme right. Below, are our louvered, 2-story barracks with the pallet-type boardwalks, which were used regularly during Monsoon season.

The hooches looked a lot like large tents, but they were actually wood framed structures that were covered and fitted with heavy olive drab colored canvas material, each of them housing three to four men. As the material make-up of these hooches afforded little protection from enemy projectiles, walls of sandbags had been put in place, surrounding all sides, except the front entry. Most of the earlier personnel lived in the hooches (including our officers & NCOs), while the newcomers, among the enlisted men, were relegated to the barracks. The former LCP (*Land Clearing Platoon*), which originally formed within 'A' Company, had quickly grown and morphed into LCT, a separate working detachment. It expanded exclusively for the clearing of vast areas of dense jungle, mainly along the region's various roadways, where heavy jungle growth had encroached significantly upon those two-lane asphalt highways. Enemy insurgency then posed a serious ongoing problem to most convoys traveling within the region. The naturally overgrown conditions allowed for the enemy's effective use of roadside cover, while launching their covert-type ambushes against military and civilian vehicles alike. The usual chain of vehicles transported men and goods to their various destinations out along these highly traveled transportation routes.

Our wooden two-story barracks were constructed of imported pine and Douglas fir (*from the U.S.*), and were rendered without any formal windows. Instead, there was a fine-mesh type metal screening applied along the front and back walls of the lower and upper levels with a window-like strip at the upper portions for limited views of the ground below, and for natural light to filter through. Multiple (*1" x 4"wide*) long slats of pine were set at a louvered angle over the remaining screened areas to allow for adequate ventilation and airflow. They also provided protection from the seasonal wind and rain, as well as nightly relief from pesky mosquitoes. As the climate was nearly always warm and humid there, the fixed airflow vents were a welcome feature and were seen as a regular part of newer building construction throughout most of the scattered military bases and Army encampments in South Vietnam (*even our mess hall had them*).

Two of our mechanics briefly posing for an after hours snapshot.

For monsoon season, a series of pallet-like boardwalks were installed all around the company area to relieve us from having to slog through the usual thick mud in the course of our daily or nightly activities. Otherwise, during the dry season, we contended with a fine dust that the beige colored soil yielded from our daily foot traffic. The dust was even heavier in the motorpool, as the plows and other vehicles regularly churned up the soil there, and ground it into a thick layer of fine powder.

At some point, I noted that some of the older Vietnamese women there had a strange little chewing habit. The use of an unusual substance called *Betelnut*, which was a small brownish-red nut from a variety of evergreen tree, called Areca, provided the user with a mild level of stimulation along with a general feeling of euphoria. However, it was highly addictive and tended to permanently stain their teeth dark red or purple, while quite amazingly resulting in the additional little-known hygienic benefit of cavity prevention. A friend I've known for years, who escaped from Hue, Vietnam and came to the U.S., recently told me that the dark stains were viewed culturally as an attractive highlight to their teeth rather than the otherwise ugly by-product of their unusual masticating habit.

At noontime, the barracks usually served as a worker's lunchroom of sorts, as many of the hooch girls gathered there at one end to spread all of their combined food offerings, cooking pans, etc., out on the plywood floor in front of them. They formed around it all, squatting in a circle while cooking and chattering non-stop in the strange cadence-sounding style of their Southeast Asian language.

When one or two of us were occasionally bedridden or restricted to light duty within the barracks, the hooch girls usually sought out a more remote spot in either the upper or lower level that was generally well removed from our bunks. All the other guys were either at the mess hall, and/or alternately working within the motorpool. Despite their courtesy in moving to another part of the barracks, I distinctly remember that nasty, permeating smell from the strange fish sauce they used as they cooked up their rice and some kind of meat and shrimp, or whatever, using small cans of Sterno as their heat source. But it never changed. They always stunk the place up with that foul-smelling stuff that almost tricked my senses the first time I smelled it, thinking there might be a dead animal somewhere nearby.

While many of our hooch girls were generally pleasant and friendly, they became more like giggly schoolgirls when they gathered together inside the barracks. They would gleefully interact in mock flirtatious chatter with the few among us there, who would sometimes persist with the occasional humorous exchange of casual friendly banter.

Many of them only spoke Vietnamese, while some others spoke a rudimentary broken form of English. (*Which I suspect, may have helped to get some of them hired by the Army in the first place*) Given that much English, we tended to understand each other fairly well, with particular difficulty often experienced in communicating some of the more detailed exchanges. When they tended to like someone, or show exceptional approval, it was usually noted with the complimentary phrase, "You numba one, G.I." Although, there didn't seem to be any mid-range numerical rating from their 1 through 10 scale, as "Numba ten" was reserved for those who may have annoyed them, or somehow caused notable disfavor. If genuine hatred was ever expressed, it pointedly came with more zeros in the form of "numba ten thou," but was rarely heard or directed at any of us. Mainly, that over-inflated number was meant for our pesky adversarial Viet Cong combatant, *'Charlie'*, or possibly even the notorious leader of the north, Ho Chi Minh.

With my own occasional attempts at playfully flirting with these hooch girls, they, in turn, took special delight in noting how young I looked at that time. I still appeared somewhat younger than my 19 years might have otherwise suggested. So they gave me a teasing nickname that did the trick in shutting me up. They called me "Cherry Boy." *If that nickname had ever gotten out to the guys, I would surely have been plagued by it for months.*

Noticing that the Vietnamese style of sitting was quite a contrast to our own, it amazed many of us that they could easily sustain long periods squatting in one spot, perfectly balanced and comfortable like that. Unlike us, they never needed a chair of any kind. As natural as anything could be, they preferred to squat down, flat-footed, with their rear end nearly touching their heels, resting unencumbered and without strain. For cheap amusement, some of us would try to imitate their type of sitting, only to quickly tire from the immediate strain on our knees, calves, and ankles. We'd usually fall off to one side onto the floor and playfully persist at trying it again with the same result.

Not entirely acquainted with the tropics, I soon found out why jungle fatigues were much preferred and universally worn. I had brought some of my stateside cotton fatigues along in my duffel bag and used them prior to receiving my full issue of poly-blend jungle wear. After wearing them a few times, I found that the all-cotton fabric tended to fall apart and rip quite easily due to the constant high humidity factor there. It then effectively reduced their further use to rags. However, with our humidity-proof jungle fatigues generally keeping us cooler in the hot, moist climate, they were also actually found to be fairly nifty. Someone designed them with pockets galore, along with side leg pouches that were plenty roomy for carrying extra amounts of small gear or even a change of socks. Camouflaged fatigues were only worn by the Marines at that time, and hadn't yet made their appearance with any of the other troops. They wouldn't until sometime after the Army's presence in Vietnam. So, pure Olive Drab (OD) was our color of destiny there…for just about everything issued.

Our accumulative dirty laundry was usually kept in a stuff-sack and left by our bunks for the hooch girls to gather up and take out for cleaning, once every week, whenever we were in for 'Stand Down'. They always did a fine job, despite the oil stains and holes from battery acid that sometimes remained, and would usually have them waiting for us the following day, pressed and folded, lying piled upon our bunks. They employed an age-old, rudimentary process in cleaning our fatigues and such: hand-scrubbing them against the rocks in a nearby stream. They followed that up with the basic drying method of airing them out, by simply exposing them to the warm rays of the sun. We would always leave payment for them under our pillows, as it was *our* responsibility to clean our clothing, not the Army's. While in the field, some laundry would be sent back by chopper, while other items would sometimes be washed and hung up in the tent to dry (*Like socks and underwear*).

Our individual living space within the barracks involved a single bunk comprised of a basic tubular metal bed frame, a simple angle iron framed, spring-link type support surface, and a 3-inch mattress. It didn't always provide for the best route to dreamland, but it still got me there. As it turned out, there was plenty of space for each man on both the upper and lower levels of the barracks, which kept us from encroaching on each other and eliminated the possible need for double bunk beds. We each had a single stall, metal locker next to our bunks as well, instead of the all-familiar footlocker that we all knew from our basic training days. We also had showers with hot water there. Our maintenance people installed a few electric water heaters, which tended to promote a more relaxed atmosphere and a sense of normalcy as we settled in each night. It was only in the field where we were often forced to compromise with a brief, cold-water shower.

Within a few months there, I advanced in rank: from Private E-2 to PFC (*private first class*). I purchased new rank pins at the PX, since many of those within the lower enlisted ranks simply wore the 'subdued' rank pins on our fatigue lapels, rather than bothering to sew on new sleeve patches. Private E-2 was the second lowest rank there was, signified by a single chevron stripe, while PFC (*E-3*) gained a lower rocker on the chevron. Private E-1 was simply the Army's entry rank, and was recognized by the complete lack of any identifiable symbol.

Private E-2 and PFC, subdued and standard type, recognized rank symbols.

In the states, our recognized rank on our sleeves was displayed in bright yellow and dark green. It was switched to shaded-black in Vietnam, and was commonly referred to as 'subdued'. All rank and other insignia worn on our jungle fatigues were required to be of the 'subdued' type, as they were less colorful, thereby less likely to be spotted in any possible enemy encounter. This fell true for all ranks in country and went without exception. Instead of the customary brass and/or silver that officers wore to distinguish their rank and branch of the Army, our officers generally wore sewn-in, fabric-type 'subdued' rank and branch insignia. They displayed them on both lapels of their jungle fatigue shirts as well as on their preferred headwear, be it an O.D. baseball cap, or a simple 'boony hat'. A triple turret-type castle served as the recognized branch insignia symbol for the Army Engineers and was displayed on one lapel to clearly identify their unit type.

The usual exchange of money there was quite different than back in the states, although the dollar value remained the same. We used our own separate currency, known as MPC, or 'Military Payment Certificates'. The bills were smaller and more colorful and graphic, with scenes commemorating the first space walk and other more recent historic moments. Denominations were much the same as our domestic currency, with twenties, tens, fives, and ones available, but since there were no coins to be used, those common denominations were also issued in smaller script (*except for pennies*).

So, when payday rolled around at the end of each month, on 'the day the eagle sh**s,' our Paymaster would always pay out in MPC notes. In the absence of script for pennies, the various merchants, along with everyone else, generally just rounded off to the nearest nickel, since pennies were absolutely useless to us. Because of black market-related counterfeiting activities, new script usually replaced the earlier bills about every 90 days or so, in an ongoing attempt to stay ahead of the game, toward effectively limiting any potential losses that counterfeiting could cause.

Front and back views of MPC script, for 10 cents and one Dollar.

South Vietnamese paper currency was in the form of 'Dong', which was the Vietnamese term for bronze. Their coinage was also called Dong, with the smaller denominational coinage referred to as Xu (*Su*). I remember it took a lot of their Dong to equal one of our Dollars, much like with the exchange rate of the Mexican Peso. In our monetary dealings with the Vietnamese, we tended to accumulate some of their currency, sort of by default. They often had only a limited supply of our currency available to them, while our unwanted acquisition of their legal tender was mainly due to making transaction change from our MPC notes. The only particular trouble with having any amount of their cash, was that it could only be used toward *their* goods and services, and wasn't accepted anywhere on post. In finding that inequity somewhat unappealing, as well as the value of some of the shoddy goods that were offered, most of the guys tended to curb some of the more frivolous buying habits there, and saved their money for items sold in the PX (*Post Exchange*).

Additionally, some amongst us had leftover stateside 'greenbacks', which were also gleefully accepted by the local Vietnamese merchants. In accepting their currency as change, one had to stay fairly updated with the exchange rates there, or take their chances with getting 'ripped off', which would happen at any opportunity that might present itself. Many of the merchants tended to test our knowledge of this, as they'd often short us on the return change, with a very sincere apology put forth in the event that one was more alert than they had anticipated.

With regard to our monthly pay, base pay was pre-determined according to rank, with everyone receiving an extra amount (*I think it was somewhere between $35 and $50*) in the form of combat pay. "Uncle Sugar" always provided a little added allowance for those of us working within a war zone. We all tended to appreciate whatever extra we could get, as working with heavy machinery and risking our lives every day in a combat zone wasn't considered to be a high salaried position in the Army. For that matter, I don't believe there were any high salaried positions, although, Chief Warrant Officer seemed to be the highest pay grade in our outfit.

~ Three ~

Land Clearing

The Land Clearing Team was normally comprised of about 65 to 70 men, each working within four platoons. Two were for the assigned tractor operators. The third served as a maintenance platoon, and the fourth contained other essential clerical personnel: Medics, cooks, etc., who helped to maintain the new company. The 'Team' also had 30 Caterpillar D-7E bulldozer-type tractors. 28 were transformed with a Rome K/G blade and a steel-reinforced, screened, open-front cab, which effectively protected the operator from most hazards while working in the 'cut'. The unusual K/G blade was angled and curved with top-mounted brush guards and a sharpened extended wedge, called a 'Stinger' that protruded from the extreme left side of the blade and was used exclusively for slicing and splitting trees. In its most complete form, this strangely modified tractor effectively cut down dense jungle growth, and was more commonly referred to as a 'Rome Plow'. Its parts were manufactured by the Rome Plow Company in Cedartown, GA (*originally based in Rome, Georgia*). The two remaining dozers among the plows were also outfitted with tree cabs, but sported standard type bull blades instead and were mainly used for certain detail work, as needed, in and around our field night defensive positions.

The Caterpillar D7-E tractor, with its inline 6-cylinder turbo-charged diesel engine, was operated through a system of dual foot brakes and hand operated steering clutches, along with a decelerator foot pedal and a 3-speed forward and 3-speed reverse type automatic transmission. Turning the dozer involved both foot and hand coordination for effective and complete left or right directional control. For example, as one would depress the left foot brake while simultaneously pulling the left steering clutch, the left track would cease movement, as the right track would continue to rotate, effectively pivoting the tractor to the left. When traveling in reverse, the opposite was true. With the left brake again depressed and the left steering clutch pulled, it would efficiently steer the tractor in that same lateral direction as it moved rearward. Right side movements were likewise performed in the same manner: by simply manipulating both the right brake pedal and right side steering clutch in harmony.

The decelerator pedal's operation was actually just the opposite from an accelerator. In this case, while one's foot is depressing the pedal with the throttle set at full, the engine will instead be near idle, as it conveniently transfers full control of the throttle to the operator's right foot. Letting off of the decelerator would simply open the throttle to whatever degree of engine speed that the operator deemed useful. This pedal gave the operator much more control of the throttle for sudden stopping and certain situations when full power wasn't always needed for the particular task at hand. Under normal field operation, the tractors would generally run at full throttle as they worked. However, without the added benefit of a foot-controlled decelerator pedal, sudden stops would have been a little more of an awkward task, comparatively. In that scenario, the operator would have had to reach up and hand-operate the throttle lever itself to lower the dozer's engine speed (*RPM*) on each particular occasion.

The angled Rome K/G blade was attached and controlled through a set of hydraulic push arms, which were mounted on the upper body of the tractor, near the engine compartment. A floor mounted (*'stick' type*) lever was positioned inside the cab for the operator's right hand control, as with most late model bulldozers. This blade, unlike others, was designed specifically for cutting heavy foliage and trees, and required a sharp cutting edge be maintained after each day's use; much like the sharp edge of a standard broadaxe. Its unusual configuration featured a prominent leading knife-edge that could be lowered no further than 6 inches from the ground. This permitted stubbles and low root structures of trees and shrubbery to remain, in the interest of preventing erosion. In that regard, it allowed for some possible future re-growth and re-generation of new foliage. Our main aim, throughout the numerous Land Clearing missions that we continually worked, was to temporarily clear the land, and not strip it.

Comparatively, a bull blade equipped dozer could only effectively push over the standing brush and trees that lie within its path, having to also ride over that downed debris to continue on. This only served to obstruct the operator's forward progress, making it somewhat tougher to continue maneuvering the tractor through the 'cut', consequently resulting in a much slower clearing process. These 'bull blade' tractors couldn't actually cut through much of anything, if left to the task, short of digging below the plant's substructure to brutally uproot things instead. But that would have made for an even slower clearing process with considerably more effort involved, and would have additionally opened up the land to erosion. Our two designated 'bull blade' dozers were also normally equipped with standard tilt cylinders. These cylinders allowed the blade to be tilted slightly to each side for optionally angling it into the soil, while moving dirt and debris in and around our Night Defensive Positions. They gave the bull blades a little more flexibility in carving out and grooming our NDP sites, as well as with the occasional establishment of LZs (*helicopter landing zones*). These notable earth-moving activities were actually more directly suited to the bull blade tractor's particular standard straight-blade configuration, as they were retained for those necessary duties, while the plows were left to work the areas of dense jungle.

Conversely, the Rome Plow's (K/G) blade was mostly fixed with a slight forward tilt being possible. It otherwise only traveled up and down, with the operator's corresponding touch of the 'stick'. However, with the proper application of this thoughtfully designed cutting tool, the Rome Plow could easily slice through heavy brush and small trees. The cut debris would mostly roll off to the right side of the angled blade and brush guards, just as it was made to do, much like how a conventional snowplow clears heavy snow from an icy roadway. While different types of plant life were encountered when operating in the jungle, the sharpened blade would often be alternately raised and lowered to whatever degree, according to plant density. In this leveraged way of cutting, the tractor could move along more efficiently, slicing through the variations of soft, as well as not so soft plant life. This in itself showcased the amazing effectiveness by which the K/G blade had demonstrated its high level of efficiency. It was found to be much more than marginally effective in clearing vast areas of thick, heavy foliage, while working in the various types of challenging terrain.

Additionally, the rear drum winch featured easily-accessed hydraulic controls at the left side of the operator's seat, just behind the shift column for the automatic transmission. This factory rear-mounted Hyster winch became a very useful and worthwhile tool in many situations, and was especially helpful during some of the more difficult days of the Monsoon season.

In further touring the inner confines of a Rome Plow, the padded black leather operator's seat was found to be a fairly roomy and comfortable enough chair, complete with armrests. It was positioned just left of center in the cab, within close reach of everything. The oil-filled hydraulic tank was mounted to the operator's right side. With a flack jacket to sit on, it occasionally became a somewhat crude 'buddy seat' for observers and instructors to use when riding along. *In those days seatbelt-type restraints hadn't been installed as a feature within any of our track driven equipment, nor within any of the unit's wheeled vehicles.*

Since LCT was, at that time, involved in their latest field operation in the Bien Hoa area, I was instructed to ride along in their supply truck and report to 2nd platoon upon arrival at their established Night Defensive Position (*NDP*). I was initially told to just hang out with the squad leader in the Deuzenhalf truck, and simply observe the operation from the edge of the 'cut' to get a better feel for this unusual type of work. So, for the first few days there, I took in the noisy, dusty scene at the edge of the jungled woodline; keenly watching our herd of tractors, as they cleared out heavily laden areas of bamboo, with some small trees and other thick shrubbery present. It was impressive, to say the least. But, after sitting around there for the suggested couple of days, I was left itching to get more acquainted with one of these rather unorthodox Rome Plow tractors. At that point, I was then finally used in relief of others, for an hour or two each day. This gave me a little 'stick time' in the cut on these strange looking dozers, and provided a little breather for a few of their grubby-looking operators, as

I got a bit of a feel for the terrain and the overall conditions there, within the ever diminishing rectangular tracts of the jungle.

Needless to say, I was already basically familiar with this piece of equipment; simply because I trained on the Caterpillar D7S series and D7E, along with the Allis Chalmer HD-16, throughout AIT, at Ft. Leonard Wood, Missouri. The old D7S was a forerunner to the 'E' series D7, and featured a small gasoline-powered 'pony engine', which was mounted on the left side of the main engine compartment. This little gas-powered engine, when cranked and running, in turn, engaged and cranked the big diesel engine as its glow plugs heated up. The D7-S also had the earlier period cable-driven type controls for its blade and winch unit, and featured a standard, 3-speed stick-shift type transmission. The newer (*1961*) D7E, along with the Allis Chalmer dozers, had all hydraulic controls and automatic transmissions that made for smoother and more efficient operation all the way around. Clearly, the old Cat 'S' series dozer, with its outdated features and controls, although effective in its own right, was an antique, compared with these more modern and efficient hydraulic tractors.

During the natural course of things, as equipment was used and evaluated, the Cat D7E emerged to be the more widely accepted 'work horse' of the Army. It was also found to be the most formidable tractor available at the time for converting into a Rome Plow. While I was indeed well acquainted with this basic bulldozer, the converted Rome Plow's unusual angled blade and steel cab were something additional to get used to. The remarkable utilization of these tractors, in this most unconventional way, wasn't at all what I had originally expected prior to arriving in country.

Within each cab, we normally kept our loaded and ready M-14 rifles stowed vertically behind the operator's seat, along with a meager supply of boxed smoke grenades (*when available*), usually stored down on the floorboard, between the seat and hydraulic tank. The smoke would come in handy whenever it became necessary to signal for any kind of immediate help, whether due to an enemy encounter or any kind of serious mechanical problem. Only one of the plow operators had the benefit of a two-way radio, for direct communication. The others could only communicate as best as possible from a distance, through the use of colored smoke, or hand signals.

Over the course of time, I had found the M-14 rifle to be the most reliable and accurate of all personally-issued weaponry there. It was also the one rifle that I had trained on and used, during my basic training at Fort Lewis. Plus, it was the one rifle which I had become so familiar with, whereby I generally came to know it part-by-part. However, there in Vietnam, it was also found to be the heaviest and longest among the different weaponry that was issued to us from time to time.

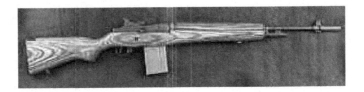

The M-14 rifle was the initial weapon issued to us and, although it proved to be quite accurate, its overall length made it somewhat awkward to readily access within the plows.

Many of these particular rifles were permanently set at semi-automatic fire, which was the norm for the standard version of this weapon. But, a few of the modified versions that were also available had a mounted switch, which could be set for either semi- or fully-automatic fire. The only drawback to its use in the fully-automatic mode was that only the first round was 'dead-on' accurate. The resulting recoil tended to send the trailing rounds above the target, where it wasted much of one's supply of ammunition. As a result, semi-automatic fire was usually more preferred.

For personal security within our cabs, there wasn't any requirement that we carry or wear our steel pots while operating in the jungle. The heavy gauge steel cab generally provided enough adequate overhead, rear, and side protection to offset the normal headgear precaution. Plus, it was often just too damn hot within those cabs to want to even consider wearing one's helmet. Instead, most of us just wore a soft O.D. baseball type cap, or a floppy 'boony' hat, or nothing at all up top. At the risk of being without our issued head protection, the persistent heat had often generated too much perspiration and discomfort to warrant the use of any kind of skullcap. But, earplugs were issued and regularly worn by all to protect against the constant threat of hearing loss. The plow's turbo-charged diesel engine usually created a considerable level of high-decibel sound when running at high RPM. Without earplugs, this condition would have otherwise caused some degree of notable damage to our eardrums over just a short period of time.

Given that all of the vehicles, plows, air compressors, and just about everything else that we used were run on diesel fuel, the smell of its exhaust was nearly always in the air. Oddly, during the dead cold winter of '67, I had come to like that odor at Fort Leonard Wood, while training on an Allis Chalmer dozer in the 'Million Dollar Hole' heavy equipment training area. It had simply become the all-identifying scent of working heavy equipment. With Land Clearing operations moving at full speed, diesel fuel was the lifeblood of our equipment. Our fuel truck drivers, along with additional tankers from our mechanized security, were kept fairly busy, driving almost daily to the nearest fuel depot to haul the precious, semi-flammable liquid back to our NDP. Naturally, this kept our field operation moving forward, and also ensured that my favorite exhaust odor would remain constant. While working in some of the more remote areas, which were usually far removed from the Army's fuel depots, Chinook helicopters would occasionally ferry 500 gallon pods of diesel out to our NDPs, as needed.

Our usual mission within the III Corp zone was a simple one, as far as understanding just what needed to be accomplished with our equipment. However, our means of carrying it out proved to be considerably more complex. The logistics in moving such a large contingent of men and equipment had to be mapped out and coordinated in accordance with our security's overall timetable of scheduled events. Initially, at the break of day, we would assemble ourselves in single-file with all of our loaded-down vehicles and trailers lined up and waiting along the road in Bearcat. Heavily laden and ready, we waited until our convoy would finally get the signal to move out, as a mechanized company of local security elements from the post set us in motion and safe-guarded our progress for much of the way.

Eventually, we would meet up with another mechanized Infantry unit along the way, as we approached our targeted work location. We'd then enter their area of operations, where they would take over and secure us throughout the new field operation. To sustain ourselves for long periods of time in the field, we had to be well stocked with all the necessary parts, goods, and ammunition that were regularly called for on each mission. Otherwise, Huey and Chinook helicopters air-lifted any additionally needed supplies into our NDP. Over the course of an Operation, we normally spent about one and a half to two months out in 'the bush', before finally accomplishing our hard earned and long-awaited objective.

Mornings arrived early and the formality of assembly was generally relaxed in the field until after breakfast. PT (*physical training*) was happily a thing of the past, for there was really no time for calisthenics in a war zone. So, all we needed to concentrate on were our plows and their pre-determined objective for the day, as well as alertly watching out for our own personal health and well-being, once our machines ventured out into the cut.

As that first operation soon ended, a new one began shortly thereafter, out in a different area of the III Corp zone. I was then paired off with a more experienced operator on plow # 28, who was a 'short-timer', soon to rotate back out to the states. He had helped to get me a bit more acquainted with the overall land-clearing process and further acclimated to the rigors of working in certain terrain and cutting trees. At first, it was a little strange, as I was like a tentative new employee, working on something that was completely different and foreign to my basic understanding of things. But, despite my initial apprehension, I quickly gained more confidence with each brief tutorial. With the veteran operator's words of encouragement, I was soon brought up to speed with all the others, and was then found to be ready to take over a plow of my own. After that, I was formally assigned to my own plow, #22, and operated it for some time before its transmission finally gave out. Then I was later switched to plow #26 because its previous operator, a tall lanky guy nicknamed 'Lurch', had also rotated out.

For the several Army Infantry divisions within the III Corp Tactical Zone, Land Clearing teams were increasingly becoming a welcome and necessary commodity. In that regard, their efforts continuously helped to further secure the region's major roadways, keeping passing convoys mostly clear of enemy ambushes, which effectively served to save lives in the process. Prior to the introduction of Land Clearing in South Vietnam, many of the major transportation arteries had been largely enveloped by extremely dense jungle growth, which commonly ran right up to the edge of these asphalt roadways. This regularly made it quite convenient for the Viet Cong to practice their hit & run type strikes, as they easily targeted our movements along the road, using the dense jungle as cover. However, with the heavy underbrush and many large trees removed, woodlines on both sides of these two-lane highways were moved back a few hundred meters, where it presented much more of a risk for enemy elements to consider the same pattern of ambush. As a result, convoys could move about more freely along the road, with less security needed. It was then somewhat heartening to note that these notorious roadside incidents and ambushes had steadily diminished in the wake of our hard earned accomplishments.

Along with the daily deployment of the plows, maintenance crews of about three to four men, normally equipped with welding generators, hand tools, and various Caterpillar parts, regularly rode the edges of the 'cut' in a couple of open bed track vehicles (*called M-548s*). They were always poised and ready in case of a break-down, or to offer assistance in the event of a belly pan fire, etc. These mechanics would also set up motorpool maintenance areas for us within our night defensive positions in the field. They were constantly working on our plows to keep them in good running order, and were always there to facilitate the needs of each operator, while helping to move the operation forward. In all of the chronicled events, regarding our outfit, not enough had been said about our amazing maintenance team. These men were actually the somewhat unsung heroes of Land Clearing, in the way they diligently worked to keep each plow operational and consistently performing within the cut. It was their great mechanical skills, ingenuity and know-how that really kept it all going on each operation, so that we could, in turn, ultimately get the job done.

The remaining members of our Land Clearing Team generally stayed in camp throughout the day, while we continued to work the plows out in the cut. In addition to their usual daily tasks, they maintained the necessary duties related to assembling and disassembling our ever-changing NDP locations. We continued to clear vast areas of jungle, as the operation steadily moved along, and routinely moved our encampments to keep pace with the operation's weekly progress. In this way, we systematically tore down and set up our NDPs as we went, until the area's clearing process was completed and the field operation finally came to an end. It became necessary to do this to keep the distance from camp at a minimum for the slow-crawling tractors, effectively cutting down on lost travel time when heading to and from each targeted area. Leapfrogging our way, from site to site, our operation's constant forward progress continued to move us ever-farther down the road, toward the ultimate completion of our mission. During these moving days, the other remaining personnel in camp would regularly take down and pack up our squad tents for us, as needed, while we worked the cut. After traveling several miles down the road, they would then unpack them from the trucks and erect them within our newly designated Night Defensive Position, to once again, serve us as proper field housing for another week to ten days. When establishing a new NDP site, each one's basic layout was always found to be quite similar to previous configurations of the usual berm encircled design, with few exceptions.

Moving from site to site, with the usual tear-down and reassembly process periodically unfolding, our existing encampment was usually alive with activity, as everyone there was keenly involved in making the necessary move. At some point, our two bull blade dozers would routinely push down the circular berm that surrounded the field compound, which they had previously pushed into place just a week or ten days prior. The field operation's primary objective usually kept the main body of plows busy, come moving day, so that our overall progress didn't have to suffer from the needed change of location. In addition to pushing down the perimeter berm, the two bull blade dozers filled in the garbage sump and smoothed everything else out within our former encampment. This restored the borrowed parcel of land to a somewhat closer semblance of its former appearance. (*Minus the heavy jungle cover, of course.*)

Headquarters Company and A Company of the 86th Engineers always provided convoy transportation support, with 10-ton tractor-trailers (*Lowboys*) designated to haul the plows. They would run us out to our initial NDP site when a field operation began, and remain back at our battalion area (*Bearcat*) while we worked. They later returned to pick us up whenever the lengthy operation ended. These 'Lowboys' even occasionally hauled us out to fairly far off locations, sometimes which were up to 100 miles (*or more*) from our main base camp. In this routinely supportive task, we would see these guys fairly often, as they regularly hauled us back and forth on all of our assigned missions. In that regard, many of us tended to consider them extended members of our overall land-clearing 'team'. Their sudden re-appearance usually signaled that all had finally been accomplished, and that it was then time to return to base camp. With 25 to 30 Ten-ton Lowboys in view, the operation would then abruptly come to a halt, and we'd methodically pack everything up and head back to Bearcat.

With the trailered plows poised to move on down the road, a few of the operators occasionally opted to ride up inside their secured tractor, myself included. We'd settle into the comfortable confines of the padded, leather operator's chair, while taking in a wide view of the surrounding countryside as we steadily rolled along the two-lane asphalt highway. This was mainly due to the driver already having a 'shotgun' rider along. With the prospect of close quarters for all three men riding abreast within the un-air conditioned cab of the truck, things would just get a bit uncomfortable at times. As we steadily followed the main route back, with others in the convoy trailing along behind in parade-like procession, our morale began to greatly improve, knowing full well that hot showers and beer would soon be available to us once again.

Not that it was altogether important, we had normally loaded our plows onto the low bed trailers front first, riding with the blade resting over the 5th wheel on the trailer's front mantled edge, just above where it connects with the 'Bobtail' 10-ton truck. It created a rather intimidating look: rolling over the countryside with the angled blade and its imposing stinger raised high in the air, as if poised to attack. However, sometime later, we were directed to load them in reverse, backing them up and onto the trailers with the lowered blade left in float position, resting down on the low bed's surface. From that point on, this became the new standard practice for loading and transporting the plows, although it left them to project a much less fearsome appearance in comparison. Our drivers always properly secured them with chains and binders, and faithfully hauled the lot of them off to their pre-determined destinations. I never really knew why or what the difference was, but, there must have been a significant reason for the change of direction on the trailer. I suspect the elevated blades might have created a little more air drag, and quite possibly may have affected the truck's overall fuel consumption. Or, perhaps the weight distribution on the trailer was better served with them secured in a reverse position.

~ Four ~

Security

Security forces for Land Clearing operations always involved a company of mechanized infantry. They were usually provided by each particular infantry division based within that province or designated area of operation, within the III Corp zone. As we moved from one divisional area to another, these highly armed mechanized units would regularly alternate their protective coverage, reflecting security for the broad general areas they represented. M-48 tank dozers, M-60 tanks, M-113 Armored Personnel Carriers, M-88 Tank Retrievers, Mine Sweepers, and other service track vehicles formed the common make-up of these mechanized security forces. They would integrate themselves and shadow us like bodyguards whenever we traveled in convoy. Plus, they would surround our targeted work areas along the jungles' edge with their tanks and APCs like protective mother bears watching over their cubs. With their usual high-level of security, they always maintained radio communication on each operation, as well as line of sight overview.

At the end of every long day, when twilight departed, the evening shroud of darkness eventually fell over our field Night Defensive Positions, as everyone there generally settled in for the night. At that time, each of the security's armored track vehicles routinely assumed their usual protective post, spread out at intervals, where they would strategically position themselves around the inner edges of the perimeter berm. These diligent men routinely stood night watch over our encampment while we worked late on our equipment, and as we slept. Although each unit's level of security varied somewhat, from our viewpoint, their constant presence was always needed and always appreciated, allowing us to relax our own guard somewhat and to focus more on our task at hand. From the 1st Infantry Division in Lai Khe to the 25th in Cu Chi, to the 11th Armored Cav Regiment out in Blackhorse, along with other notables within the III Corp zone, they all took special care to safeguard the efforts of our Land Clearing missions. In turn, it greatly served their own cause, since the telling effects from our labor naturally brought about an ease in security to each newly cleared sector, along with the important benefit of gaining more control over their general areas of operation.

On one operation, we drew a mechanized 'Aussie' unit and found those boys to be a good, tight, field security force, and a real fun-loving group of guys. Several of them would regularly gather with us at night for some lively conversation and a few good laughs. One particular morning, after they departed our NDP for the last time, another 'yank' mechanized unit took over control, as we awoke to find them gone. In their absence, and much to our amusement, we soon discovered little painted (*white*) kangaroo mementos, stenciled onto the sides of each of our plows.

On another operation, an ARVN unit (*commonly referred to as Arvins*) was assigned to secure us. They were a South Vietnamese Army mechanized unit, seemingly without a lot of past experience in securing others. Along with their mechanized company came an Infantry company of foot soldiers, who were engaged in daily patrols in and around the general outlying area. Many of us felt a strange uneasiness with them watching over us, and were actually glad to see them leave when the time finally came. In this case, perhaps there may have been a little pre-judged over-reaction on our part. But, it was a common assertion amongst us that many of these guys lacked the staying power and commitment needed to be effective in securing combat support units. We had always heard that when the going got tough, some ARVNs would desert. Fortunately, no major incidents occurred while in their care to give any further credence to our general suspicion and uneasiness, although many sighs of relief accompanied their eventual departure.

All of the plows were numbered in order to keep track, and know just who was who as they moved about each day. In addition to numbering the sides and rear surfaces of the cabs, large white numbers were painted on the tops of each to easily identify the machines from the air. Our brass also allowed us to personalize our plows, if we so desired, by painting names on the sides and rear of the cabs. It sort of created a close, personal identification with the equipment, which had a somewhat positive effect on most of the operators. The names often reflected an operator's nickname, or where he was from, or what he may have enjoyed, or perhaps even a pun of sorts. Some of the names were: 'Kentucky Plow Boy', 'Soulful Strut', 'The Pusher', and 'Sugar Bear'. I mostly operated plow # 22, early on, and named it, the 'Sopwith Pickle', which was a strange little twist on the 'Peanuts' comic strip, as Snoopy's imaginary WWI bi-plane was a 'Sopwith Camel'... and since my plow was green... oh well, what can I say? My feeble attempts at humor, at the time, were a bit skewed or strange and none of the other guys really understood it either, if there actually was much of anything to understand about it.

Our assigned mechanized security had the daily chore of watching over us, as we diligently worked toward leveling the war's 'playing field'.

Along with the names on the plows, many of the operators had been given personal nicknames, some colorful and some not so colorful. At times, a few of the guy's last names were either too hard to remember or too strange to pronounce, which usually gave rise to coining a suitable nickname. While some just came by their particular social handle through more natural means, other nicknames tended to border on something humorous. Commonly, first names weren't used much in the Army, unless, of course, while occasionally in the company of a few close friends. Otherwise, each man was typically called by his last name, unless an appropriate nickname had been coined. Some of these more celebrated nicknames within our outfit were, 'Ski', 'Tuck', 'Squirrel', 'Lurch', 'Preacher', 'Cornbread', and 'Sugar Bear', among others. In my case, I really didn't get a nickname, or at least one that I knew of, since my surname, 'Brown', which amounted to only one syllable, was easy enough (*to roll off the tongue*) as it was...I suppose.

While clearing away jungle from along the various roadways that most of our field operations focused on, the daily scenario usually involved around 14 Rome Plows (*depending if all were operational*). They were normally relegated to one side of the highway (*Or, in some cases, both sides*), and each set of 14 was led by a 'Lead Plow'. The Lead Plow operators, who were generally the more experienced among us, carried a PRC-25 radio with a helmeted headset in order to maintain a COM line with a small, 'OH-6,' Army observation helicopter overhead. Our company commander or platoon leader usually presided there, over the initial outline of the cut, or *'Trace'*, as it was called. From his aerial overview, he would instruct the

Lead Plow on degrees of turn, when guiding and drawing out the primary rectangular outline of the area to be cut. The plows always addressed the cut in a counter-clockwise direction, as the severed trees and brush cleanly rolled off to the right side of the angled K/G blade, away from the plow's path, just like it was designed to do. Traveling clockwise instead would have simply defeated the whole purpose. Piling the cut debris to the inside of the trace would eventually just serve to bog down all the other tractors there. Those trailing the Lead plow, within the cut, simply followed behind as it made its straight line cuts to complete the outline of each targeted tract. In this way, they would all effectively cut down the remaining jungle growth and trees inside the established trace. Following each other around at different intervals, they cut concentrically toward the middle of the tract until it was entirely leveled, before progressing on to the next appointed sector.

The number of plows that were used tended to vary, depending largely on what was required to complete the mission and on what was to be cut, along with the number of machines that were up and running at the time.

In maintaining our overall security each day, our assigned mechanized infantry forces would always follow us into a newly cut trace after it was initially established by the Lead plow and the first wave of followers. Moving along at intervals, they would generally surround the area to effectively stand guard over us as we worked. Periodically, they would move in closer toward the middle of the tract, as the foliated area would progressively diminish in size from our collective activity there. While watching over the plows each day, they remained poised atop their tanks and APCs, to respond immediately to any possible enemy activity thereabouts that might suddenly occur.

Each plow operator knew his job and, for the most part, also knew what to expect in performing it. If any of us had become slack in our daily duties, it was quickly pointed out with a face-to-face encounter from a squad leader or our platoon sergeant. In the event that an appointed operator just couldn't handle the task of operating and maintaining a Rome Plow, for whatever reason, he was usually sent back to our company area in Bearcat (*or, later in Long Binh*), and replaced with someone who possibly could. However, most of us were eagerly up to the task, learning as we went and dedicating ourselves to the cause. In that regard, we willingly endured the rigors and hardships that naturally went along with it.

While working with the plows every day, it was a little interesting to note that, although each one was nearly identical in appearance, they all felt strangely different in some way. They often showed little subtle differences, or even quite pronounced ones, in performance and operational control. This was always discovered, as a few of us would move from plow to plow on occasion. The newer ones, of course, tended to perform better with fewer mechanical problems occurring, though feeling somewhat stiff compared to the older and more seasoned tractors. However, some of the older ones were plagued with chronic breakdowns, leaving operators with more than their fair share of frustration. But, as we operated and maintained our machines on a daily basis, we came to know them well and were even occasionally able to detect certain small mechanical problems occurring. Oftentimes, we managed to catch these problems just before they further developed into a more serious breakdown.

In pressing to establish the trace, the Lead Plow would often encounter different levels of enemy activity within the freshly-cut areas of dense jungle. Many of those who had assumed this 'Lead' position had been known, from time to time, to sustain damage and occasional injury as a result of these sudden violent encounters. It was simply an unavoidable hazard, since the Lead Plow was the 'spearhead' of each initial cut, having to work in advance of the security by blazing the basic trail for each targeted area, or tract. When operating as Lead Plow, the view was somewhat limited, and oftentimes one could not see much of anything except for the heavy brush and trees that the tractor was cutting as it moved blindly through the jungle. But, this designated 'initial assault' type Rome Plow always maintained a dedicated straight-line cut under the reliable radio guidance from above. The only actual deviation from this straight-line cut was a brief adjustment to avoid an occasional large tree, or a bomb crater that might lie within the plow's otherwise appointed path. Because visibility was often very limited in the Lead Plow, it was mainly left to those in the chopper above, to routinely keep a topographical map aboard in order to point out any significant obstacles over the radio.

Typical limited view, from the Lead Plow position.
(*This photo was taken when I assumed the Lead Plow role*)

Somewhat later in my tour, I found myself working the cut in the Lead Plow role. On one such occasion, our CO had to call for me to abort the trace because I was taking in small arms fire that he had a 'visual' on from his overhead perch in the 'chopper'. From my position within the thick bamboo and other dense under-brush, visibility was very limited; oftentimes, I just couldn't see much of anything beyond the immediate periphery of the plow. My radio headset on, combined with the constant din from the roaring engine, made other sounds minimal at best, and not always recognizable. Without a clear view, I simply continued to plunge my way through the endless sea of tall, dusty bamboo stalks that lay in my immediate path. At the time, I had heard some unusual pinging sounds but they never really registered as anything out of the ordinary until our CO blasted me with a verbal: to "get the hell out of there! You're under fire!" With the sudden realization of what was actually happening finally getting through to me, I instantly put the transmission into third gear reverse. Then, I let off of the decelerator pedal and raised the blade to nearly full height, before quickly slumping down onto the floorboard of the dozer. At that point, I didn't really care where it might go; just somewhere away from there and out of range from the AK-47 fire that suddenly had my heart pounding blood at a much more accelerated rate. As I emerged from the interior of the bamboo jungle, the other plows had alertly observed my rapid retreat and seemingly understood what was happening up front. Instinctively, they too, pulled back from their trailing positions and quickly followed me out of the immediate area.

With the plows bunched up a safe distance away, our security's tanks and APCs quickly converged on the area, to effectively clear out the hot spot (*like the cavalry coming to the rescue*). This allowed us to eventually regroup and continue on. In their haste, they turned a corner and rushed past us, as one tank collided, in a glancing blow, with one of our other plows down the line. Observing things while sitting it out, I noticed one of their M-60 tanks had actually overshot the established pathway as it hurried along, and struck the side of the halted plow, where it wound up breaking a few links in the dozer's track. Amazingly, the unharmed tank simply bounced off and continued on as if nothing much had occurred.

While we continuously worked our plows out around the cut, there were some major incidents and accidents that occurred from time to time. A few good men were lost or injured as a result of these occasional unfortunate events. Enemy contact within the trace was, again, an unavoidable hazard, as well as a nagging reminder to each and every plow operator who had actively participated in this risky business of Land Clearing. Small arms fire (*AK-47*), RPG's (*Rocket Propelled Grenades*), and B-40 rockets were the usual ordinance that was sometimes fired at us, as we continued to cut down portions of the enemy's vast blanket of overgrown protective cover. We simply tried to concentrate on our work, as enemy guerrillas would try to interrupt the process with potentially deadly force, of one type or another. Land mines were also an ongoing problem and a few of our plows sustained some degree of damage when those incidents occurred, only occasionally injuring an operator. These mines varied in size, most being the small, anti-personnel type that didn't inflict much damage at all to the heavy tractors. Although some of the larger ones certainly did, to some extent, with the operator subsequently suffering whatever degree of consequence that tended to accompany the blast.

In an effort to help minimize the effects of land mines and to further protect our operators, each plow and bull blade tractor had been modified with heavy gauge, steel plate type, mine guards. These were installed at the entry to each side of the cab to help deflect most potential incoming shrapnel. The mostly enclosed, heavy steel reinforced tree cabs provided a good measure of protection to the operators as well. At night, mortars and rockets were also occasionally fired into our Night Defensive Positions, mainly to rattle us and steal our precious sleep (*or, so it seemed*). Through it all, we were usually pretty fortunate, in that only few casualties ever occurred among our guys from the occasional late night harassment.

Falling in line behind the Lead Plow and cutting out the remainder of the tract day in and day out tended to become monotonous to many of us, when working in predominately flat terrain. Given this particular view, along with the constant lulling drone of the engine, some of us occasionally drifted off mentally and fell asleep at the controls for a brief time or two. Because most of us regularly put in long hours during the day, along with additional time for night maintenance, the amount of restorative sleep that we got often fell a bit short of the norm. It left some of us feeling somewhat tired as we worked, occasionally setting us up as easy prey for the daytime 'Sandman'. I remember when it happened to me on a particular occurrence. My squad leader noticed me dozing when I came around near his position, so he ventured out into the cut area and picked up a big dirt clod. Throwing it my way, it accurately hit the side screen on the cab. Its sudden impact exploded the clod and rained its fine dirt in on me, rudely waking me and getting my full and immediate attention.

Not surprisingly, I would resort to the very same tactic nearly a year later on another operator, when I later stepped off the plow to become a squad leader.

Miles of similar-looking jungle and unseen hazards that lie ahead made remaining alert extremely critical to each operator. Being attentive and able to quickly react to sudden situations was crucial, since the enemy notoriously specialized in surprise hit-and-run type tactics and the planting of various types of land mines that were inevitably encountered on each operation, regardless of terrain. Plus, there was always the nagging thought that an RPG with one's name on it might be lying poised on a launcher, somewhere out within the thick underbrush of the jungle. However, the ongoing monotony and occasional lack of sleep sometimes pulled some of us away from our otherwise guarded approach. I admit that, from time to time, my attention drifted into daydreams a bit. In repeated trips around the trace, I would occasionally become hypnotically involved with various thoughts of home, and other idle fascinations that came to mind. On those occasions, while falling into a semi-trance-like state, I would sometimes catch myself drifting off into the slumber zone. It was easy to do since the same type of brushy, low-level terrain was plainly, still in full view. With earplugs firmly in place, the softened constant drone of the engine also contributed to lulling one away from alertness. An occasional tap on the left brake and a pull on the left steering clutch when turning the corner on the cut was about the only notable activity that drew the operator's attention within these dull areas of brushy terrain. But, for the most part, I kept a fairly vigilant watch over things. Having only received enemy fire on two separate occasions, I had also impacted my plow on a few land mines over the course of time. Fortunately, they were all of the 'anti-personnel' variety, which caused little or no damage to the undercarriage, nor to either one of the plow's tracks.

These were the very anti-personnel mines that I had earlier dreaded when joining the Army. They had apparently been placed in anticipation of ground forces moving through the area, not Rome Plows. However, much larger versions, in the form of 500 lb bombs, were sometimes laid out within our appointed path, directly targeting our tractors. The sudden, earth-shaking impact from one of these larger landmines occasionally caused significant physical damage to a few of them, with moderate to serious injury to the operators. Given the few expected enemy encounters that I experienced, I believe I was really quite a bit luckier than some of the other guys in the outfit, somehow steering myself clear from sprays of AK-47 fire, while also avoiding a few rockets, and potential shrapnel from the mines that I occasionally ran over. Several of our operators received various wounds, whether just minor lacerations of the skin or significantly more serious deep shrapnel-type wounds that required some degree of hospitalization. Looking at things in retrospect, had I instead been drafted to likely become an unprotected 'Grunt' infantryman, no doubt, the odds would have weighed more heavily against me in assuring my survivability from that war. So, it gave me special reflective pleasure in knowing that I had apparently made the right alternate choice, by purposely placing most of that heavy, protective steel securely under my boots.

In cutting out these large tracts of dense jungle, we all tended to set our own pace as we worked. For whatever reason, some of the plows often moved along at a little more accelerated rate than a few of the others. Occasionally it even became necessary to pass some the slower plows in the cut in order to further maintain one's established pace. We also tended to cut around a stalled or broken down plow to keep it from interfering with the flow of traffic, where it might otherwise slow down and bunch up the rest of the pack. Efficiency depended largely on each operator and how well they were able to move about through the thickly-wooded landscape without encountering too much in the way of halting problems. These problems included getting hung up on a stump, or breaking a hydraulic line, among others. If the radiator happened to overheat, the operator had no choice but to stop and let it cool down while adding more water. The other plows would then simply continue on around him, until the machine's cooling system was restored to the proper temperature range. Of course, with each downed tractor, either our platoon sergeant or a squad leader arrived on the scene to inquire as to the exact nature of the halting problem. While supervising the activity in the cut, it was naturally always their desire to maintain the mobility of the equipment, and try to keep things moving on schedule in any way possible. With the usual daily mechanical problems and breakdowns, our M-548 maintenance tracks would frequently be called to the scene to offer assistance and immediately repair any light damage or mechanical failure that may have occurred. Otherwise, they would remain within the general vicinity to observe the operation as we worked, where they'd readily avail themselves to our particular incidental needs.

Since our plow platoons ordinarily worked separately from each other, some small competitive challenges had developed at some point. In establishing these competitive drives, one platoon would effectively cut out a bit more acreage than the other to gain special recognition and thereby achieve their own little morale boost for the week. It gave their egos good reason to believe that they might actually be the more productive of our two Rome Plow platoons (*A third plow platoon was added sometime later*). This was just something to make things a little more interesting for the operators and, at the same time, was a subtle little ploy to raise the overall rate of jungle demolition for the operation. However, pushing the plows to the limit for the sake of a higher rate of cut also tended to produce more breakdowns. The burden of these competitive drives then tended to rest a little heavier on the shoulders of our mechanics and their ability to maintain a large enough stock of replacement parts. Additionally, any downtime suffered by a more serious breakdown then tended to reflect more negatively on the production rate of cut, whereby it would keep the downed tractor sidelined and unproductive until the necessary repairs could be made.

As time went by, safety then began to take precedence within the cut, thereby lessening the overall importance of doing a speedy job each day. As a result, our brass and NCOs then started to focus more on ways to cut down on accidents and potential enemy fire, while allowing for a more normalized rate of cut. In applying a measure of forward intelligence to our otherwise short supply of advanced knowledge, reconnaissance planes and observation helicopters were commonly used to fly over the areas which were targeted for clearing. This was done to check on the terrain and overall conditions, along with any potential enemy activity. At the same time, it coordinated things well in advance of the plows' impending deployment into that particular targeted sector. These 'fly-overs' tended to give our officers a fuller overview of things in these areas, as they gained a little more knowledge about some of the particular hazards that might await us. At times, they were even able to spot signs of enemy activity, seeing the VC moving about below. When this occurred, air strikes or artillery fire would commonly be called in to clear out much of their presence, prior to our active involvement there.

On one such occasion, I watched two (*what appeared to be*) Navy jets swoop down to about 100 feet from the ground. They delivered their load of phosphorus bombs to their target, just a short distance across the way from where we were working, and departed the area nearly as quick as they had arrived. The exploding phosphorus lit up the area and sparkled tremendously, as the blasts sent up bright incendiary clouds that soon subsided and diminished into billowy clouds of white smoke. The delayed thunderous 'boom' from each impressive blast hit my ears just a few seconds later. It was quite a sight to see — and a welcome one at that — to watch our 'flyboys' in action., especially whenever it pertained to our own safe working conditions within the cut. When it came to these types of enemy encounters, they were sometimes called upon at a moment's notice, to give us a helping hand with our occasional embattled situations. At other times, napalm was dropped over some of these areas, which also had its own deadly incendiary effects on all forms of life that happened to lie within its appointed path.

When these types of payload carrying aircraft weren't within range of our work areas, Huey 'Cobra' helicopter gunships were occasionally called in. They had mounted rockets along with rapid machine gun fire available to take care of pocket area hot spots of enemy activity. These newer 'Cobras' were sleek, modern style helicopters that were built for speed and maneuverability when delivering their 70 mm rocket-type ordinance and 20 mm 'Gatling-style' cannon fire to their target. Their much sleeker appearance featured a forward nose cockpit, with a second cockpit running in tandem (*just behind it*), much like the inline cockpits on fighter jets.

Another quite notable helping hand that was called out on rare occasions of increased enemy activity, during our land clearing missions, was an assault plane nicknamed, "Puff, The Magic Dragon." It was actually an old Air Force C-47 transport plane that had been upgraded and converted into an air-assault plane, featuring 7.62 mm mounted mini-guns that could astoundingly fire 6,000 rounds per minute. Generally, they had three of these amazing mini-guns mounted on one side of the aircraft. It had the overall potential of 18,000 rounds per minute when all were put to use; or, one bullet per square foot within a football field sized area in less than three seconds. This proved to be quite devastating firepower that cleared out enemy infiltrated areas, by virtue of near complete annihilation.

While mowing down the amazingly thick jungle growth with our Rome Plows, along III Corp's main serpentine transportation links, enemy tunnel complexes were occasionally discovered. These often revealed rice and weapons caches, which in many instances yielded some rather impressive amounts.

Enemy combatants were also sometimes found inside these tunnels. When this was detected, members of our security would have to go in, conduct a search, and then flush them out or kill them if they resisted. Those unique individuals, who were small enough and brave enough to accomplish this, were known as 'Tunnel Rats'. Armed with a flashlight in one hand and a .45 cal. pistol in the other, these men had the unenviable task of searching the narrow tunnels to extract any and all important contents, including *'Charlie'*. Eventually, they planted explosive charges to destroy these subterranean havens, which effectively prevented any possible reuse by other enemy elements.

Sometimes, these meticulously carved out tunnel systems were quite extensive. Large living quarters and rudimentary kitchen facilities were established down there, along with adjoining tunnels that spanned out all around the immediate area. They were usually well designed, methodically dug out earthen labyrinths that could still, surprisingly, support the heavy weight of the working tractors overhead without incurring cave-ins. I'm sure that some of these amazing tunnel complexes were never exposed, despite our thorough clearing of the jungle above them. Although, when their occupants finally did emerge from below, they must have been shocked and disturbed by the absolute loss of cover that dramatically changed the formerly wooded landscape. It undoubtedly appeared more like an open wasteland to them, compared to the protective wooded sanctuary that had formerly helped to serve them in their nearly invisible form of opposition.

In an attempt to silence our opposition and give us a little sanctuary for a few nights, it was proposed that we set up a temporary NDP within a sizeable graveyard that we had come upon, just as one field operation began near the area of Trang Bang. It was a ploy that seemed to work for us, until the following night when we were hit with heavier than usual mortar fire. The sporadic rounds came whistling in on us and we were forced to take cover wherever we could in our short-term encampment, which was situated just immediately adjacent to this headstone-lined graveyard. I happened to be in the vicinity of the garbage sump, at the time when the attack occurred, just casually talking with one of the other guys. We both suddenly froze as we heard the first ones come in, and looked at each other in utter disbelief. We then quickly jumped down into the 10-12-foot deep sump to momentarily escape the aerial threat. As the first wave of mortars subsided, we scrambled out to head toward better overhead cover. After that shaky experience, we moved our encampment and soon set up a properly bermed Night Defensive Position, which was thankfully nowhere near any kind of graveyard or shrine.

At night, in addition to guarding the perimeter area outside the berm, our security would send out Long Range Patrols. These small reconnaissance teams established 'listening posts' away from the NDP to report any sign of enemy activity in the vicinity. They would normally stay out there all night and return in the morning. It was a standard security measure that was regularly employed in the field, which tended to offer some additional information on enemy movements to our mechanized protectors. For our immediate perimeter security, outward visibility at night was improved with illumination flares that were shot up into the air, as needed, where they would slowly float to the ground by way of tiny parachutes. These provided a level of bright orangish-yellow light for a few minutes over nearby areas where movement may have been detected.

While we tried to sleep, our security's APCs and tanks were commonly spread out all around the inner edges of the berm. Each one's occupants attentively watched and communicated, in an around the clock effort to keep us all from harm. We generally got accustomed to the random popping sound from illumination flares being fired off, but never really got entirely comfortable with the unexpected audible announcement of 'INCOMING!' when mortars and/or rockets were suddenly being fired in on us. Comparatively, it wasn't all too dissimilar from the sound of someone yelling, "FIRE!" in a crowded movie theater, except for the actual sight of mortars or rockets exploding nearby.

Normally, when under attack at night, we would try to make a mad dash to the nearest sand-bagged bunker, sometimes tripping over tent ropes along the way. I always found that the best place to seek refuge and ensure my safety was under my plow. I had to mostly lay flat on the ground after diving under the rear of the undercarriage, since there wasn't a lot of overhead space for sitting up. If a rocket or mortar happened to strike the plow, it still wasn't likely to rain anything in on me down there. Otherwise, the area behind the blade offered fair protection and room to sit and wait things out. When used in this way, many of us felt a greater sense of safety, as the plows not only offered their solid steel protection to us while out within the cut, but also during the evening hours, where they further served as reliable makeshift bunkers. For this very purpose, we often parked them close-by, fanned out in a protective arc just behind our tents.

When establishing our more basic needs, portable single-stall outhouses were brought into the NDPs for us via a Lowboy truck. As for safe drinking water, potable water was trucked or helicoptered in, as needed, in metal water tanks that were mounted on small trailers equipped with spigots. Huey helicopters would commonly ferry other needed items along with incoming and outgoing mail, miscellaneous parts, and occasionally new replacement personnel. In turn, a few 'short timers', on the verge of rotating out sometimes jump on the outbound chopper to ultimately catch that always-popular aerial icon, the 'Freedom Bird', when finally making their way back home to the states.

As a rule, we usually kept one or two bull blade dozers in camp for carving out LZs (*helicopter landing zones*), and for pushing up the protective berms surrounding our night defensive positions. They were also needed for our own effective method of trash disposal. Within our NDPs, 10- to 12-foot deep sumps had to be dug for this purpose; and, at some point, these garbage sumps would usually be doused with diesel fuel and set ablaze before finally covering them over when departing the area. The distinctive smell of burning plastic sometimes filled the air from it, which helped to create a quasi-nauseous effect on some of us. Human waste would simply be buried, since we would only remain at each NDP site for a brief period of 7 to 10 days, depending on our plows' expeditious forward progress.

On one unusual moving day, we began the routine of breaking camp in the morning, before pressing on toward the next, soon-to-be-established NDP site. Meanwhile, the two bull blade dozers were busy pushing over and flattening out the large circular berm that had formerly surrounded our field compound. At about the same time, several miles away, our plows were carving out the jungle-covered area where our next NDP site would soon be re-established, along with its adjoining acreage. Upon later arrival of these bull blade dozers' at this new encampment, the two were offloaded from the lowboy trailers and the operators received general instructions on laying out the new perimeter berm. They then began to routinely push up a fresh 5- to 6-foot high circular earthen border around the newly established site, just as they had ordinarily done, numerous times in the past.

Within the passage of a day or two there, some of the men detected a distinct foul odor coming from one part of the berm. After unearthing the suspect section, we discovered two bodies that had been inadvertently buried there. Without a proper explanation for this, it could only be deduced, as to how it may have possibly come about. It oddly appeared that a tunnel must have been inadvertently positioned right under where part of the berm was to be formed. These two apparent Viet Cong guerillas must have been exiting just as a plow was nearly on top of them, while simply clearing out the existing wooded acreage there. It also seems that the operator was likely unaware of their presence there, being fully absorbed in the activity of cutting the jungle down when this strange mishap came about. One particular grisly detail from this weird incident was that one of the victims was missing the top of his head, which could only be attributed to the sharpened leading edge that was the prominent slicing feature on the plow's K/G blade. Of course, this all went completely unnoticed by whichever of our plow operators who was working in that particular area when it occurred. Sometime later that day, the unnamed bull blade operator, who was working on establishing that part of the berm, in an effort to complete our standard circular perimeter, then proceeded to establish it right where those two had met with their untimely deaths. It was quite apparent that he also hadn't noticed their mutilated and trampled bodies scattered there, as he methodically worked his tractor in the usual manner. He simply pushed up a large volume of dirt and unknowingly covered them over, in the course of routinely establishing our common circular earthen fence line.

After finding these two VC within the berm, the security quickly rifled through their pockets and took what they had before carrying them off and burying them. It was quite an odd way for us to get two enemy kills, but very effective, never-the-less; even though there was no way of knowing which one of us had unintentionally accomplished the feat.

Somehow, I had always found it to be extremely odd, if not unlikely, that these circumstances actually came about in this way. But, the fact of the matter remains, that the two bodies were recovered from within the berm, and no one was forthcoming in any way to take credit for it, or to knowingly explain just how that might have occurred.

Leaving our old NDP, on another operation, we had only advanced about a quarter mile up a narrow dirt access road toward the main highway before an M-60 tank up in front of the convoy hit a sizeable land mine. I caught sight of the sudden fiery explosion while riding in the back of our open-bed Deuzenhalf truck, in the company of a few others there. While casually looking up toward the front of the convoy to view the road ahead, I suddenly caught sight of the blast, and could clearly see the 6 or 7 Infantrymen who were riding atop the tank, sailing through the air like rag dolls. Almost immediately, I felt the surprising wind-like force of the blast and shuddered at the thunderous sound that quickly followed. It was a rather large explosion, which sent up a plume of grayish black smoke and appeared to cause serious damage to the big tank. Several medics and a few others rushed in to help the wounded, who were found strewn about the area there. At that point, all vehicles had stopped. We were then all looking like confused 'sitting ducks' until it was soon determined that the convoy should continue on and go around the tank. Quite sadly, it's grim wreckage just stood there smoldering, with its occupants lying nearby, injured and dazed.

While the downed men were being attended to, we moved out again in an attempt to reach the main highway, and further lessen the chance of an ambush occurring somewhere out along this narrow access road. In view of this shocking development, some of the other APCs and tanks remained with the damaged tank to keep the wounded well secured there until Med-Evac choppers could arrive on the scene to properly expedite them to field hospitals. While continuing around the severely damaged tank, we could see that the driver was still down in his seat, screaming in pain. It was later related to us that the blast had broken both of his legs and an arm, so he couldn't extricate himself as he was. Otherwise, he was alive, as were all of the others, but visibly out of his mind in fear and pain. The wounded men on the ground were left dazed and seriously disoriented from the blast's concussion, while nearly all had sustained broken bones and/or deep lacerations from the resulting shrapnel. With the sobering display of wounded and wreckage scattered about, this distressing scene strangely unfolded at that point into a more disturbingly surreal moment. It had clearly become a stunning eyeful for all of us on the truck, as we slowly passed by the big tank. While unable to offer our own help, we knew all too well that we simply had to keep moving.

Just a short distance beyond that point, we stopped again and I was unexpectedly recruited, along with another man on the Deuzenhalf, to walk ahead and carefully inspect the surfaces of the single-lane dirt access road for any other possible land mines that could still lie within our path. We were instructed to occasionally probe the ground within the suspicious earthen tire paths, using a bayonet, since we didn't have any minesweeping gear available to us for this apparent sacrificial act. This sudden exercise was really appalling to me. I had never been trained to do this sort of thing, and immediately felt threatened by the thought of possibly coming in contact with that which I had dreaded from even the time before my enlistment, in the late summer of '67. So, although not knowing exactly what to recognize, we resigned ourselves to the task and nervously walked beside each other in slow-motion, carefully inspecting while lightly stabbing the ground as we went, to possibly detect any solid metallic object that the rest of the convoy might otherwise avoid. The other vehicles and APCs slowly followed, some distance behind us, until we got somewhat closer to the main highway, where the probing was finally called off, and we both relaxed and breathed two huge sighs of relief.

~ Five ~

Field Maintenance

In our daily grind with the plows, we always worked farmer's hours, going from sun-up to sun-down. At the end of each days' venture out in the cut, upon returning to our Night Defensive Position, we would then regularly work on night maintenance, as needed, for oil and filter changes, blade sharpening (*with pneumatic grinders*), winch repair, etc., to keep the plows and our two bull blade dozers fit and ready for the following day. Also as part of our regular daily maintenance, grease fittings were filled and fuel tanks were topped off with diesel. Additionally, air filters and the entire engine compartment and cab were blown out with a powerful blast of air, using a hose from one of our trailered diesel-powered air compressors. With these compressor engines roaring at high RPM, causing the accumulated air-blown dust to swirl about, we would methodically blow out everything in sight. It was really the only way to effectively remove the day's heavy build-up of dust and leafy debris from within the cab and engine compartment, as well as from the core of each tractor's radiator. We would also turn the air hoses on each other, given that we were all just as dusty and dirty as our machines had become. Since we didn't have protective masks for this regular chore, rags were often used and worn, much in the same way that earlier period cowboys utilized their simple bandanas.

The large, drum-shaped air filter element, which was located within a similarly shaped metal housing, where it was plainly mounted on the right side of the engine compartment, also had to be removed and thoroughly blown out each day. Its usual, thick accumulation of dust from each day's activity was always significant. If anyone had ever neglected to blow out this air filter, additional dust would surely have built up on it to where it would gradually cause blockage and affect the tractor's overall performance. Their plow would then soon suffer a significant loss in power, whereby it would ultimately stall and become inoperable due to low airflow to the turbo-charged engine.

This necessary NDP maintenance event, being the final phase of the daily clearing process, was largely just routine activity for us. It regularly repeated each night after we brought the tractors in from the cut, and got some chow. With our maintenance mechanics and squad leaders there to help monitor things throughout the evening, they helped to ensure that most everything got accomplished, and that each plow was revitalized and ready for the following day.

The nightly motorpool scene was usually lit up and awash with various maintenance activities. A series of lights were often set up there to give us just enough illumination, aiding our nocturnal efforts as we worked on the tractors. Amid the assorted maintenance tasks there, our hand-held, pneumatic grinders regularly shot steady streams of sparks into the night air as we put new edges on our blades for the next days' challenge. While grinding a new edge with our blades elevated to about three feet, one had to be extra careful not to brush against the newly honed blade because it was usually ground down to become nearly as sharp as a kitchen knife. Bright intermittent flashes from our mechanics' arc welders also strobed the evening scene. In these instances, they often made random repairs to damaged cabs, brush guards, and whatever else that might have needed further hardened reinforcement on the heavy steel plows. Occasionally, one of the K/G blades would somehow lose a short section of its cutting edge, which may have broken off while working in rocky ground, or possibly while cutting and lifting against a particularly tough hardwood tree. Our maintenance mechanics would then fashion a suitable section of steel that closely matched up with the missing piece, and then artfully weld it into place. When that was solidly welded in place, they'd grind it down to match the surrounding edge's bevel, where it nearly restored the blade to its former, factory honed, knife-like cutting surface.

Whenever it was deemed necessary during night maintenance, belly pans also had to be cleaned out. They were unbolted on one side, allowing the heavy hinged pan to swing down for relatively easy cleaning. They were then re-attached using a large pry bar to hold it back up in place, while the pan's bolt holes were re-aligned with a small, pointed bar (*roller-head pry bar*), before it was re-bolted to the tractor's undercarriage. It ordinarily took at least two men to address these steel pans, though, since they tended to be much too heavy for one man to perform the task by himself. New replacement parts for the plows and the company's other various type track and wheeled vehicles were always kept in a large utility van near the motorpool area, which was our main source for nuts & bolts, gaskets, 'O'rings, and other essential parts whenever more involved maintenance was called for.

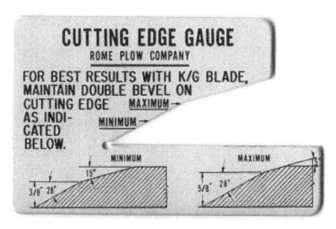

This was a metal blade gauge that was issued to all the operators. It was from the Rome Plow Company, of Cedartown, GA, and came with an accompanying storage envelope. As an accessory item, it was mostly found to be somewhat of a novelty piece, and in that regard, was rarely used. Despite its technical details, we were usually well enough acquainted with what was called for, to keep these blades sharp and cutting efficiently. The following instructions were copied here, from the gauge's envelope:

    CUTTING EDGE GAUGE - FOR ROME K/G BLADES

    For efficient operation of the K/G blade, it is important to keep the cutting edge sharp and the proper shape. Use the gauge to check the shape as the cutting edge is being sharpened, keeping in mind these important points.

1. MAINTAIN SHAPE: A double bevel provides the best results - a long bevel at the rear and a shorter bevel at the front edge. If the front bevel is flatter than indicated, the edge may chip or roll. If it is steeper than indicated, unnecessary power will be used for shearing. The long bevel is required to reduce the blade thickness and wedge action as the blade slices through a tree. Sharpen by regrinding front bevel only until the maximum front bevel depth is reached; then regrind both bevels to obtain the shape with minimum front bevel.

2. SHARPNESS AT BOTTOM: If the underside wears (*caused by having the blade tilted forward too much*), a sharp edge cannot be obtained in the proper place. Even if the edge feels sharp, it will tear and uproot rather than shear.

3. REGRIND TO PROPER SHAPE: The portion of the edge that is worn away on the underside, must be ground off and the cutting edge reground to its original shape.

At day's end, showers were certainly a more than welcome necessity, as well as a long awaited relief for the lot of us. For much of the day, we quite literally wore the thickly accrued layers of dust from the jungle on all of the exposed parts of our bodies. During the dry season, each plow operator had to endure the heat of the day, in addition to the heat from the big diesel engine that further raised temperatures within the open front cab. This compounded condition made the hot days feel even hotter within our cabs. In that particular region of the world, it was not altogether unusual for temperatures to occasionally soar somewhere between 110 to 115 degrees (*Fahrenheit*). To our dismay, we found ourselves receiving our own fair share of these almost unbearably hot days. As this intense heat regularly mixed with the high humidity factor there, we all tended to sweat profusely and found that, as daylight departed, the dust which was kicked up from our daytime activity had built up and caked thickly on our skin over the long course of the day. Before hitting the showers, we often appeared as dark as night, or not unlike the appearance of many working West Virginia coal miners. When blowing our noses, the accumulated mucous appeared black as well, due to breathing in all that swirling dust that we stirred up on the exposed floor of the jungle. We didn't have dust masks available for this, but some of us eventually learned to keep a rag nearby to breathe through when the dust momentarily got exceptionally thick. Eye protection wasn't provided either and small particles from the kicked-up debris would occasionally lodge in one eye, or both. On those occasions, it was sometimes necessary to stop and flush things out with a good dousing of canteen water. Otherwise, each day's accumulative effects from the jungle were pretty much like those of the previous days' involvement.

As these long days would finally come to pass, and as we continued to encounter the usual (*and unusual*) jungle-related conditions, we learned to overcome many of the daily annoyances and hardships that went along with the clearing process, and pushed on with our efforts, to eventually finish up the cutting, and ultimately accomplish the mission. To simply say that every day out there was an adventure would certainly be a fair statement to make, but wouldn't be very revealing nor encompassing within the general scope our daily escapades.

For those of us who preferred hot or warm showers, the remaining hot water from the plow's radiators was often drained off into five-gallon cans and poured into one's overhead makeshift canvas (*Australian*) shower bucket (*for those operators who had one*). This clever bit of resourcefulness helped to keep some among us a little happier and more at ease in the field. Otherwise, some of us were resigned to take a quick cold shower, since any shower was certainly better than no shower at all. It wasn't exactly cold, though, just on the cool side. That was because the region's tropically warm outside temperatures tended to remain fairly constant.

Amazingly, as we adapted to life within the company's NDP sites, our steel pots were found to be quite versatile, where they occasionally came in rather handy. The 'pot' had effectively served a variety of uses in the field, least of which was to actually wear it. Some among us would even shave on occasion, using them as basins after removing their outer cloth cover, whenever time permitted. Using it as a wash basin, it also provided an easy means to clean a few pairs of socks, whenever needed. We also found it to be an excellent cooking pot, as it was sometimes put to use with a can of 'Sterno', whenever the moment called for something hot and unusual to satisfy a late night hunger craving.

After completing one's night maintenance and hitting the showers, there often wasn't much additional time for anything else except sleep. Maintenance would keep many of us occupied well into the night, depending on the overall condition of each operator's plow.

After retiring from the duties of each long day, we then took our turn on 'COM watch' throughout the night in 2-hour shifts. We were required to remain awake during those periods in our squad tents to monitor the radio, while writing letters or reading books or magazines. Essentially, it was a form of guard duty within each squad tent to keep us apprised of the nightly activities in the area. This was a way by which we could remain on 'alert' and not become too surprised by a sudden rocket or mortar attack, or even perhaps a potential enemy ground assault. With the airwaves becoming a sort of early warning system, we listened through the radio's headphones to the chatter from our mechanized security elements, reporting from all areas around the berm. In the morning, everyone would be awakened by the last one on watch, and we would all go about our normal business in preparation for the new day's repeat performance of the ongoing land clearing process. But, of course, on occasion, someone would fall asleep during their time on watch, to defeat the whole purpose of the exercise. As exhausted as most of us were, from the arduous activities of the previous day, this sometimes couldn't be helped.

Flack jackets were regularly provided for each operator to give some additional upper body protection from land mines, etc. However, not everyone wore them. The heat and humidity, again, was sometimes overwhelming and the plow's steel cabs naturally conducted the heat of the day, along with the engine's generous contribution. Most operators would simply go shirtless and douse themselves with canteen water during those hotter periods. Even giving our pants a good soaking with water offered some level of welcome relief from the overall hellish discomfort that generally accompanied the dry season.

Our thick, steel 'tree cabs' always kept us well shielded and shaded from the rays of the sun throughout the day, but the intense heat which built up on the metal's surface simply transferred through, making the cab stiflingly hot at times. Because of the dry season's naturally extreme heat factor, combined with the penetrating heat from the engine, it often got terribly hot and seemingly unbearable at times. I remember that the exterior surface areas of the cabs were sometimes so hot that we learned not to grab onto the bars or screens when entering or exiting during these times of extreme heat. This unwanted heat of the day gradually increased during the noon hour, as the Sun's intense rays began to bear down on us. Fortunately, these hotter periods usually only occurred from around

1300 to 1600 hours (*1 p.m. to 4 p.m.*), during the warmer months there, giving us somewhat milder conditions throughout the morning.

Sometimes while out around the cut during the afternoon hours, I occasionally found myself looking down at my watch, in hopes that the Sun's punishing heat might finally be on the wane, and cooler conditions would only be minutes away. The late afternoon relief from those intense solar rays was always noticeably significant.

The heavy duty, rubber hydraulic lines (*which were attached to both upper ends of the plow's push arms*) tended to absorb heat from the hot pressurized oil within and, mixed with the additional heat given off by the engine and searing sunlight, they periodically needed replacement, short of cracking and rupturing somewhere out along the cut. Once in a while, one of the lines would also break from rare accidental contact with some of the cut debris. Generally, when this occurred, an occasional branch would weave its way up and suddenly impact on one of the lines, causing it to burst the pressurized hose, where it resulted in an unexpected shower of hot hydraulic oil, with no way of anticipating the direction of its pressurized flow. As a result of this, every so often, a few unlucky operators received one of these surprise 'oil baths.'

Fortunately, most of us were able to get out of the way when some of the spray entered the cab, and burns tended to be minimal, if at all suffered. Typically, it left the hood and engine compartment fouled with oil, with the hot smoke stack sizzling and smoking additionally from it. The spilled oil sometimes covered the floorboard in the cab, and often accumulated on one or both tracks. The operator then had to take extra care when exiting the cab, to cautiously grab hold of the outside screen, while carefully trying to step down from either side to avoid any possible injury from slipping on the oil fouled surfaces.

I remember when I took a partial 'oil bath' the first time I experienced a blown hydraulic line. The floorboard was so slippery with oil that I had to grab onto anything I could find that was unaffected in order to ever-so-slowly inch my way out over the mine guard, as I awkwardly maintained my balance. Then, while desperately clutching the heavy steel mesh screen on the side of the cab, I gently lowered myself down onto the equally slippery track, where I could carefully ease out to the edge of it before letting go to safely jump down onto the ground. This took a little time, but if it wasn't done slowly and methodically, just like that, I would surely have otherwise slipped and fallen onto the steel track below, to then suffer some unknown consequential fate. When suddenly experiencing this annoying slippery condition, it became virtually impossible to move about freely on the plow's solid steel surfaces once they had become fouled with oil in this terribly inconvenient sort of way.

Another particularly bothersome thing associated with these messy incidents, in the event that an operator happened to receive a full or even partial 'oil bath', was the remaining fact that there wasn't a change of fatigues available when something like this occurred. When our clothing became soaked with oil, we still had to continue on and just live with it, until eventually returning to our NDP at day's end to finally shed those fouled, oily threads, and desperately hit the showers.

Whenever that sudden 'oil bath' event occurred, one's plow could not be operated again until the broken hydraulic line was replaced and the system was fully restored and cleaned up. In most cases, the interior of the cab was in desperate need of a thorough cleaning as well, prior to bringing it back on line with the other tractors. In anticipation, our maintenance crews in the field always kept a good supply of replacement hoses and additional hydraulic fluid aboard the M-548 maintenance tracks, which often stood by and helped to monitor the movement of the plows as they worked the cut. They always replaced the broken hydraulic line on site, even while the other plows worked the cut around them. Afterwards, they would get the machine thoroughly cleaned up, to where it was presentable enough to again resume normal activity with the rest of the plows.

This was a somewhat typical motorpool scene within one of our field Night Defensive Positions, with an air compressor at center stage, and the plow's tree bars lifted, in order to pull the radiators out for cleaning, and to afford more room, so our maintenance mechanics could more easily work on the diesel engines. Note, from this view of the backside of the Rome Plow, the rear drum winch sits prominently below the fuel tank, and the accompanying right side mounted hydraulic tank.

When a particular situation called for a special effort in restoring a 'downed' plow, our maintenance crews would usually meet that challenge while helping to maintain our daily rate of cut, sometimes being required to work well into the night. At times, they even worked into the morning of the following day before finishing their task, in a conscientious response to fully repair and restore any partially operable plow. In applying their skills, they usually got it fixed and back on line in relatively short order. On these occasions, tracks were sometimes disassembled, repaired, and reassembled. Engines were partially rebuilt, and transmissions were even repaired or replaced, in rare instances (*depending on component part availability*). Additionally, hydraulic systems were also disassembled and revamped in the field, with new replacement parts and seals. All of these not-so-easy, time-consuming tasks usually came with a targeted deadline to meet, come morning.

When major parts were scarce in the field and a plow couldn't continue on without them, the tractor was usually loaded onto a lowboy trailer and hauled back to our main motorpool in Bearcat (*or later, in Long Binh*). There, a rear echelon maintenance team would usually make the needed repairs, depending on availability of the particular parts in question. If the repairs could be accomplished while there was still sufficient time remaining on the current field operation, the plow would then simply be trucked back out, to resume normal activity in the cut. This was also true whenever one of the tractors was wrecked by enemy contact, whether it was from an RPG or a land mine that was significantly larger than the common anti-personnel variety. In many cases, our heroic maintenance crews regularly performed some minor miracles on these partially wrecked plows, restoring several of them that otherwise might have been regarded along the same lines as 'Humpty Dumpty' after his fall. Others that were damaged too severely were simply known as 'dead plows', and were usually left abandoned somewhere out in the company motorpool, where they could later be cannibalized for any of their remaining salvageable parts.

~ Six ~

Monsoon

During Monsoon season, the constant torrential rainfall regularly made for a muddy mess within our NDP sites. The daily activity of the tractors churned up the saturated soil, creating deep, sloppy areas of mud within the motorpool area, as well as along the established access trails that led in and out of these provisional field encampments. While our trucks found it next to impossible to gain enough traction in this creamy brown slop, oftentimes, a plow or a bull blade had to be called upon to pull them through it. In applying oneself to this, the machine's tracks sometimes disappeared under the deeper sections of liquefied standing mud. Trying to get around on foot through this mud was also a challenge, with an occasional slip and fall into the slick gooey mire. We just didn't have any type of solid surface relief from it, like our boardwalk pallets back at Bearcat had provided us with. During these wetter periods, we tended to keep fairly dry in our tents, sleeping above ground on foldable canvas cots. Keeping our feet dry during this seasonal soggy period was an even further challenge to us, as our jungle boots were not waterproof and dry socks were sometimes in short supply. Jungle Rot was a common foot ailment that some of us contracted and endured during these extreme wet weather months, and for sometime thereafter. It was a particularly bothersome fungal type skin condition, which was brought on by not adequately keeping our feet dry enough to prevent its tell-tale and somewhat debilitating affects. A few of the men, including myself, occasionally had it so bad we were sent back to our battalion area for treatment, along with some time off for it to properly heal. It would itch like crazy, with multiple open sores appearing on most areas of each foot, as many of these exacerbated lesions often advanced up past the ankles. It was a particular type of skin condition that just couldn't be remedied in the field, where the ideal prospect of keeping one's feet completely dry had always remained somewhat slim during the rainy season.

The season of Monsoon also tended to make things considerably more difficult for the plows, as we continued to cut through the variations of jungle in different terrains and affected soils. When encountering woodland type jungle, the footing was found to be generally good. The remaining thick root structures, from the severed vegetation, generally maintained a fairly firm surface, which enabled the heavy tractors to gain solid enough footholds as they worked, despite the nearly incessant rains that thoroughly dominated a lot of the days of Monsoon. Within these areas of woodland type jungle, there were also some irregular open areas that caused only a few halting problems, with regard to getting the tractors momentarily stuck in the somewhat shallow brown mud. Although, when in Bamboo, which proved to have a much shallower and less extensive root structure, the mud tended to run deep in spots, causing our plows to bog down more often and get helplessly stuck in its gray, quicksand-like goo.

A fellow operator posing for the camera, while awaiting help with his plow deeply mired within the sandy, gray mud, after attempting to cut through the typically heavy thickets of bamboo that tended to dominate the landscape in some areas.

In this type of mud, the plows would commonly get so bogged down in the liquefied mire, to the point where their tracks could no longer be seen. They would just churn and go nowhere, except to sink deeper within the quagmire there, while the futility of the operator's efforts to escape from it was soon realized in this sand-based soil that remained so prevalent in areas of thick bamboo. Fortunately, for these near helpless situations, each plow was equipped with a hydraulic-driven rear drum winch, which often proved to be a very useful and needed implement for restoring the plows' footing, and getting them and the operation back on track.

When help wasn't immediately around, and when a sizeable tree might have been close-by, a mud-stuck operator could always help himself by securing his winch cable around the tree to serve as an anchor, and let the winch simply pull the plow out by winding the cable up on the drum, while at the same time, keeping the transmission in neutral. As the tree normally wouldn't budge, the plow certainly would.

This particular type of mud was somewhat different, in that it tended to be more liquefied and gooey. Its rather slick composition tended to prevent many of these track vehicles from gaining any serious bite. The mud also stuck heavily to our boots and pant legs, as we often had to trudge through it to attach our winch cables to trees, or to other tractors, APCs, tanks, or even some trucks on occasion. Sometimes during that necessary cable hook-up, a few of us even managed to get a leg stuck in the deeper areas of thick mud. We tended to struggle for a brief period, before finally pulling it free and getting on with securing the cable. The worst part was to get a leg stuck and lose one's balance while trying to get it free, only to fall back into the sticky goo; or, to be forced to just sit down in it before finally wresting the leg free and moving on. As it was, many of these little frustrating incidents were often expressed in an expletive-laced display of seething exasperation. At other times, the shear helplessness of the situation instead, just brought about uncontrollable, hysterical laughter in simply giving in to the utter ridiculousness of it all.

I recall, on a very wet and soggy day, one of our security's M-60 tanks had gotten deeply stuck in the thick gray mud. It took two Rome Plows to eventually pull it out and get it back to stable ground. A single plow initially attempted the task, but, as these large tanks tend to carry a lot of weight, it simply bogged the plow down and got it stuck as well. In the heavy downpours that normally came with the Monsoon season, our daily progress had slowed a bit as we struggled with the extreme muddy conditions there. But, despite the conditions, we managed to slog on through it and were able to eventually clear all of the targeted terrain that the operation had originally called for.

Around 2 a.m. on a particularly stormy and wet early morning, the relentless pounding of wind and rain finally broke the moorings loose on my squad tent, collapsing it onto us as we slept (*there were about 6 of us in it*). We frantically scrambled to get out in order to regain ourselves, before having to fight the elements, and properly re-erect the tent. After struggling with it for some time, while battling the fierce winds and driving rain, we finally managed to get it back up and re-secured. However, the by-products of our joint heroic efforts were our completely wet and muddy clothes, and our fully expressed frustrations with the elements. While it wasn't funny at the time, we somehow later managed to laugh about these sorts of storm induced mishaps. The weather monster of Monsoon may have soaked us to the bone on occasion, but it never really succeeded in dampening our spirits all that much. Despite the extreme conditions and the little incidental inconveniences that we were forced to endure, overall, our morale remained pretty much on an even keel. For the most part, everyone pulled together while living and working in the field, like the highly cooperative team of engineers that we were.

During daylight hours in Monsoon season, temperatures generally remained comfortably warm. At night, only a mild coolness would settle over our encampments, as humidity levels generally remained fairly high year-round. Even when it wasn't raining, it never got all that cold there, so we never had any real cause to pack our heavy field jackets. However, needless to say, we got plenty of use out of our rain gear, which essentially amounted to a hooded 'poncho'. These ponchos were somewhat unwieldy in their design and awkward to wear while working, so string was often used to wrap the dysfunctional rain-gear close to our bodies in order to work on the equipment without getting the vinyl material hung up on anything around the engine compartment or elsewhere.

While operating in the jungle, as the rain often came down in heavy downpours, it sometimes became difficult to see the terrain ahead. Under those conditions, visibility was often limited, and the way ahead just became a big blur up in front of the blade. But, with another plow in partial view up front, the cut line was still found to be somewhat decipherable, and we would simply persevere and continue on. However, depending on the rate of rainfall and type of terrain, it sometimes became necessary to pull all of the tractors out of the cut, for safety concerns, and wait until the stormy conditions and persistent deluges subsided somewhat before resuming our ongoing attack against the enemy's amazingly thick, but temporary domain of cover.

Despite the irregular presence of mud, cutting jungle in wet weather conditions was actually a much more comfortable experience for the operators, in that the heavy dust from the jungle floor was virtually absent during Monsoon season. Comparatively, it was such a joy to breath in the moist fresh air, and actually pick up some of the sweet scents that the jungle's plants and flowers gave off as we steadily crawled our way through the cut each day. Also, the otherwise extreme solar heat factor was mercifully absent during the rainy season, which allowed the plows to perform much better. This seasonal reprieve effectively helped to cut down on many heat-related equipment break downs that tended to be more prevalent during the warmer months. Additionally, it gave us all a pleasant escape from the daily tormenting 'furnace blasts' of the dry season, while offering our sweat glands a long-awaited, and much deserved respite.

~ Seven ~

Chow

As squad tents were used in the field to house groups of 6 to 8 men, one was also used to establish a mess tent, where we would take hot meals at breakfast and dinner. For this, we had our own cooks working daily to provide us with good, nourishing and mostly satisfying Army chow. It wasn't anything to write home about, as no one ever got overly excited about powdered eggs in the morning, or powdered milk, or even the notorious military standard, 'Shit on a Shingle'. But, bacon, sausage, and ham were always prominent staples on our breakfast trays, and helped somewhat in flavoring the powdered eggs, or so it seemed. Fresh baked breads and biscuits, along with plenty of tasty meat and potatoes and green vegetables were regularly offered up on the evening meal, with fruit-impregnated Jell-O and fudge brownies sometimes included on the menu as well. There were also a variety of beverages available to us: from coffee, tea, orange juice and apple juice in the morning, to iced tea and Kool-Aid at dinner. Kool-Aid actually became more of a common beverage there, alongside iced tea, as it could be stored indefinitely, and was very easy to make.

For these morning and evening cooked meals, we all generally enjoyed what was offered to us, for the most part. When presenting our food trays, ample portions were always available for those big appetites that our work created. It's just that sometimes the apple pie and ice cream spilled over onto the mashed potatoes and gravy, or drifted into the stew to make for quite a different sort of taste altogether. As a result of the server's careless efforts, where they oftentimes heaped large portions of food upon our trays, it was also fairly easy to realize that beef gravy wasn't all that savory when combined with chocolate cake, as well as the few other encountered variations of unintentional culinary incompatibility that we occasionally had to put up with. But, despite those little annoyances, we actually ate well enough and tended to enjoy most of our cook's daily offerings.

At Thanksgiving, all the essential faire of turkey and the trimmings were well presented to us, as we celebrated the traditional holiday as best we could under the circumstances. Despite the fact that we were thousands of miles from our stateside homes, and deeply embedded within a war zone, many of us still had much to be thankful for. This Southeast Asian rendition of our national holiday's traditional offerings had actually helped to further raise our spirits a bit, as we came together to recognize that day. Also, every once in a while, these cooks would satisfy our home-felt cravings for some good old All-American comfort food, like hot dogs and cheeseburgers with French fries, along with fresh baked apple pie. Knowing how variety tended to spark interest, our mess sergeant and cooks generally composed a somewhat diverse assortment of main dishes for the week. Along with their usual accompaniments, a different dessert was offered most days, so that it wasn't always the same old thing. In time, as our field operations progressed and our food supplies got a little on the lean side, we tended to see some of the more common offerings a little more often than we did when we started out.

When finished with 'chow', we were sometimes left to clean our own food trays out in the field. For this, a few 55-gallon drums of water were placed just aside from the mess area for a quick wash and rinse process. A couple of these drums had kerosene heaters attached to provide hot water for proper cleaning of the trays. After we scraped our excess food into the trash bins, the trays and flatware were then dipped and cleaned with a G.I. brush in a drum containing hot soapy water, and dipped again to rinse in cold water before finally plunging them into clean hot water and placing them in a stack. Most of us didn't mind the little added chore, as the cooks took pretty good care of our daily caloric needs. When they had doughnuts, I would smuggle a few onboard my plow in the morning and chomp on them when I could. When steaks were available, they'd grill 'em up the way each of us preferred them, in an effort to keep us well contented with their various culinary skills. Their regular gastronomic contributions were always well noted and greatly appreciated by most, if not all, within our fairly close-knit company of men. In keeping us all rather well fed, their daily offerings also helped to maintain our overall health and strength, which allowed us to further handle the constant rigors and trappings of Land Clearing.

In the morning, we usually filled our canteens at the water trailer (*or sometimes in the mess tent, with juice*), and some of us would take a few warm cans of soda along in the cab, since a warm soda still seemed a mite better than no soda at all, to satisfy one's craving for sugar. But, with the combined heat factors in the plows always creating a fairly powerful thirst, a couple of canteens of water were maintained daily, until refills were eventually called for. At the start of each new day, our platoon's Deuzenhalf truck usually accompanied the plows out to the cut, towing a water trailer behind for us to refill our canteens from time to time. At some point, salt tablets were handed out to everyone, as the hotter days of the dry season came increasingly more into play. There was no dallying after breakfast. We all knew the importance of being ready to depart for the trace when the order was given. Being late was simply inexcusable.

Every day after breakfast, the operators were all issued C-rations to be taken onboard the plows for the lunch meal. At mid-day, they would regularly be waved aside from the cut to form up in a semi-circle and shut the engines down. We'd take our noon break and eat lunch right there in the cab, while everyone else associated with working the cut did the same. These cardboard boxed C-rations weren't all that good, but were certainly nourishing and at least halfway satisfying. I remember some of the main entrées, which varied from Salisbury steak to beef stew, or spaghetti and meatballs, among others. These main course selections always came packed in olive-drab colored cans, as did the fruit and dessert items, all of which required the use of a tiny P-38 military can opener. The canned pound cake proved to be a rather tasty and somewhat popular dessert item in itself, to which some of us occasionally maintained our own personal collections of these government issued, semi-confectionary delights (*For some of us, they were often held in reserve, for trading value*). Otherwise, the remaining items within the individual box of C's came in a plastic packet, containing things like crackers, butter, jam, plastic utensils, towelettes, a mini-pack of four cigarettes, a matchbook, and a packet of instant coffee, along with some sugar and powdered creamer. Occasionally, some of us saved the canned main entrée along with packets of saltine crackers, and kept them in reserve in our tents, where, with the use of our steel pot (*helmet*) and a can of 'Sterno', one could cook up a late night snack. As for me, I usually gobbled down my lunch fairly quick since the break was often a good opportunity to shut my eyes and take a relaxing, but brief cat-nap for a few more minutes before re-cranking the engine and continuing on out around the cut.

Trading rations with our security was also a fairly common practice. It made for a welcome change from the rather mundane C-rations, which were the exact same types of rations used in Korea and in the latter part of WWII. These mechanized security units usually carried LRP (*Long Range Patrol*) rations, which were dehydrated pouches of spaghetti and meatballs, or beef stew, or lasagna, etc. Surprisingly, they were actually much better than expected and not at all unlike the dehydrated or freeze-dried hiking and camping food pouches that are readily available in sporting goods stores these days. Fortunately, in pursuing our yen for something different, some of the mechanized infantry boys, in turn, also regarded our C-rations as an equally welcome change from their own somewhat ordinary 'eats.'

Almost every day was a work day while on these rather lengthy field operations. As each day came and went, oftentimes one could easily lose track of which day of the week it actually was, or even the calendar-numbered day of the month. Sunday tended to bring it all back, though, as our Chaplain regularly conducted Catholic and Protestant services that day at dawn in an open area of the NDP. Unless one was of a different religion entirely, our CO usually frowned on inattendance, in case anyone chose to skip out on the services. He also encouraged letter writing, as 'mail call' always provided an additional little boost to company morale, especially when one would get a 'care package' from home. If you had gotten homemade brownies or chocolate chip cookies while in the field, they generally wouldn't last very long, as everyone pretty much knew that you couldn't realistically keep them all to yourself.

My mother had written quite regularly throughout my tour, and my little sister sent letters and a care package from time to time, along with a few photos that some of the other guys enjoyed, asking if they might write to her themselves. Writing, at that time, seemed to be somewhat of a chore for me, especially since I hadn't mastered my letter writing skills as yet. I was mostly 'spent' at the end of each day anyway. But, when I did write home, I typically kept things fairly brief, although somewhat upbeat, and portrayed my daily activities simply as hard work, which it was. They didn't need to know about all the other little details… not then, anyway.

As our field operations progressed, each operator was occasionally granted a day off within the NDP, to rest up, write letters, and generally take care of personal gear and the like. These rare, peaceful breaks were a welcome relief from the usual daily gritty grind that went along with operating a plow, as well as from the regular nightly motorpool maintenance that was additionally called for at the end of each day. Plus, just staying clean for one day was indeed a pleasant change, and a real joy in itself.

Since we never had a barber along with us in the field, the lot of us appeared somewhat shaggy and unshaven much of the time. Our daily focus was largely on the plows, with little time to devote to our own personal maintenance. Oftentimes, we would remain in the field for periods of 45 days or more, depending on the amount of territory that was to be leveled, and the overall conditions that came with it. About mid-way through each operation, the lot of us would begin to appear more like a rag-tag scurvy bunch, while taking on an un-shaven and un-cut appearance that clearly went against Army regulation. My platoon even picked up a nickname that reflected on this: we were known as 'Brennan's Animals' for a while, until platoon sergeant Brennan finally rotated out. But, as long as the plows were well maintained and our performances met the operation's expectations, our brass and higher ranking NCOs generally made allowances for our somewhat unkempt appearance, and would wait until we returned to Battalion HQ (*Bearcat or Long Binh*), while in on Stand Down, to ensure that we all, once again, measured up to Army standards.

The only way in which we received updated news reports about things back in the States or elsewhere while out in the field, was from a few among us who had battery-operated am/fm radios along. While in our NDP's, we were usually much too busy to pay attention to anything news-wise, that might have been going on in the world at the time. But, as we settled into our tents at night, we could get news and music broadcasted over the am band from AFVN, which was the Armed Forces Radio Network in Vietnam. They would play the latest hit songs of Rock 'n Roll, Country, Soul and Jazz and report the top U.S. news on the hour, along with some 'in country' military news. But, mainly, we were interested in hearing a little familiar music as we wound down from the day's stressful activities and poured ourselves into our squad tents for the night. From time to time, we also heard reports about a gal on another station, who was loosely referred to as, 'Hanoi Hanna'. She was apparently the North Vietnamese's version of 'Tokyo Rose', and often chided our soldiers with her pointed misinformation, to: seriously consider their loved ones, and abandon the foolish war effort against the Viet Cong and the communist Army from the north. I never actually heard any of her nightly communist propaganda, even though the AFVN radio reports of her broadcasts often sounded quite amusing.

In other local happenings, I remember a cameraman and field reporter from CBS News came by our NDP to get a story on Land Clearing operations while we were working in the 1st Infantry Division's area of operations, just a little north of Di An (*pronounced Dee-on*). While I was still in camp with my plow being worked on, the reporter came by and asked me and a few of the mechanics some questions on the plow's general operation. A little later, when the plow was cleared to go, I was told that this same reporter was interested in riding along with me for a short distance to get a general feel for how these machines perform and maneuver. As it was, I was overdue to catch up with the rest of the plows and re-join the cut. But, I soon relented, and put my flack jacket up on the hydraulic tank while showing him to his crude seat, there within the cab. After instructing him to brace himself and to "hang on to something", I then trotted my plow a short distance out of the NDP, where I intended to brazenly show off a little.

Getting him acquainted with the plow's general maneuverability, I turned a quick left, and then a right. I noticed him bouncing and struggling to stay seated, as he tried to brace himself the best he could, all the while knowing that he had little available to hang on to. After making a sudden 360-degree turn in second gear, he then waved at me to stop. Apparently, he had had enough. From the look on his face, it appeared that he was close to losing his lunch, as it was probably just a little more than he had expected. I guess he got his story, though.

Meanwhile, back in the states, a political firestorm had been raging against the current administration, as the build-up of negative public opinion about the war continued to feed those flames. 1968 was a Presidential election year, and anti-war protests were regularly seen all around the country. It was the final year for President Lyndon Johnson, since he refused to run for another term, while openly showing signs of frustration and despair. Having become the main target of protesters, he had not succeeded in steering the war toward its ultimate victory, for which we had involved ourselves to begin with. He wound up taking the brunt of public opinion for our continued involvement there, and was then seemingly unwilling to push toward the original goal, as he was unable to re-unite a decidedly divided Congress. So, instead of advancing to take the fight directly to the enemy, we remained largely garrisoned, as an occupying force, to mostly let the enemy bring the fight to us. With that, the war had then become more of a protracted involvement with no real end in sight. Our overall efforts had then become more of a slow, measured success, as we were still succeeding in our encounters with the enemy, to a large degree, but with significant loss of life.

However, now the war had come home to the states. While many back there vehemently disagreed with our continued involvement in Vietnam, most of the men in my unit continued to honor our presence there, just as I did, as we pressed on with our daily activities. In all appearances, the next President would ultimately inherit the task of pushing on toward victory, or relenting to the ever-mounting, media driven anti-war movement. So, as it was then, we just didn't quite know how everything was to be played out toward reaching our original objective, in this ongoing struggle that greatly affected the established four-zone elongated landmass of South Vietnam.

As time rolled on, 1968 proved to be a rather big news year both here and back in the states. Within that same year, a crazed gunman felled Martin Luther King, while Democratic presidential candidate, Bobby Kennedy likewise met his untimely demise at the hands of an assassin. The news from home never seemed to be all that good, as political unrest and violent death always appeared to dominate the airwaves. We only had the medium of radio available to us for our regular daily rendezvous with news and sports back in the 'real world'. However, it was indeed fascinating to hear that pitcher Denny McLain had become major league baseball's first 30-game winner since the 'dead ball' era. In accomplishing that feat, he helped to lead the Detroit Tigers to a World Series championship that year.

Continuing on in our field support operations, we cut through a lot of thickly-grown vegetation within the different divisional territories of the III Corp zone. Through our relentless and persistent hard work, we helped to secure many more vital roadways in the process. Wherever we happened to go, we were warmly welcomed by the mechanized infantry elements there, since we were ultimately making their jobs somewhat easier by greatly improving the overall security conditions within those enemy infiltrated areas.

On a rare occasion in the field, while we worked around the 25th Infantry Division's area near Cu Chi, as Nui Ba Den (*Black Virgin Mountain*) loomed off in the distance, I was stuck again in camp with my plow needing a part. While working on it with an assist from a couple of our mechanics, I noticed a tractor-trailer rig in the distance, winding along the dusty road below, which led all the way up and into our bermed compound, its long plume of dust trailing behind it. As it finally approached and entered the compound, the truck proceeded right through the maintenance area and parked with the rear of its boxed trailer facing us. When the dust cloud finally cleared, the driver got out and came around to open up the double rear doors on this 40-foot refrigerated van, then jumped up inside and hollered out like a carnival barker, "Alright, who wants Ice Cream? I've got Chocolate, Strawberry, and Vanilla!" He tossed down the frozen half-gallon cartons to the 10 or 12 men that just happened to be on the scene then, and distributed plastic spoons to us all.

As it turned out, this truck was sent out by the Commander of the 25th Infantry Division in appreciation for our efforts in clearing the heavy foliage along his division's roadways and around particular enemy area strongholds. Our jungle clearing activities there significantly reduced the amount of men ordinarily lost to roadside ambushes and area incursions, which made him ever so grateful. Needless to say, our mess sergeant had more than enough ice cream to add to our evening meals for the next few days that followed.

I ate nearly the entire half-gallon carton of strawberry that I got and enjoyed it like never before, as I sat down on the ground like a school kid with my back up against a truck tire. It was a luxury food item that we never had while in the field. Beer happened to be another; although it became widely known that many of our officers had their own private 'stashes' of the liquid amber, including our CO. There was indeed a little bit of a 'double standard' applied when it came to alcohol in the field. However, quite often in the Army, that's where rank tended to have its privileges.

~ Eight ~

Misfortunes of War

During the course of our field operations, we traveled pretty much all over the III Corp area. Moving out in convoy, we went as far south as Tay Ninh, as far east as Xuan Loc, as far north as Song Be, and as far west as the 'Iron Triangle' area, and the Cambodian border, with occasional overlaps into the IV Corp area also occurring from time to time. Each operation was different, as far as terrain and types of vegetation. We sometimes cut a lot of bamboo, while, at other times, thick woodland type jungles were encountered, with heavy underbrush, along with a few occasional groves of rubber trees and some tall, thick 'Savannah' grass areas as well.

In a way, it was somewhat sad and seemed such a shame that we had to destroy so many thousands of acres of prolific, thriving plant life, along with the disruption of tropical wildlife habitats within these targeted areas of the jungle. But we knew that it was for the greater good. In the interest of better security for all who traveled the highways there, we tended to establish cleared swaths of 200 to 300 meters on each side of every road that was encountered, and generally maintained that set pattern for those troubled areas that had been slated for demolition. In further serving the security interests of our Infantry forces, our efforts also provided them safer access into enemy stronghold areas, where our usual activities tended to greatly diminish the enemy's continued presence there. In certain instances during these more difficult times of war, some extreme measures were occasionally needed and employed in order to enact more viable solutions to equally extreme problems. Mechanized land clearing, in this case, was called on in Vietnam to fill the vacancy of that void, and to help counteract much of the enemy's out of control insurgency.

Cutting Bamboo

In observing the vegetation that was encountered within the various areas of III Corp, it was easy to note that there were likely hundreds of different vibrant plant species growing within the jungle, intertwined and competing for space, largely without any established theme or order. In many areas, it was mainly a mix of thick, overcrowded, random growth, with an obvious rich terra firma that generously fed their root structures, nurturing them to become such notably ample and well-developed plants. Large, flower-laden vines had often curled around trees and other shrubbery, stretching the plant's tentacles out far and wide in a nice, colorful floral display of trumpeted blooms. It seemed to suggest that its potential for prolific growth was almost limitless there within these naturally exotic botanical gardens, which comprised just a small portion of the general make-up within the vast areas of jungle that we regularly encountered.

Unfortunately, the reality of war forced us to disregard the natural beauty and majestic wonder, which abounded within these amazingly immense areas of thick tropical forests. The local Viet Cong, along with the invading NVA forces, had considered the normally dense high-canopied overgrowth to be their own exclusive haven and sanctuary. In that regard, they effectively used its cover against us while practicing their persistent hit-and-run warfare. To deal with this problem, our forces could find no other way to counteract the situation than to remove enough of the enemy's cover, via the Rome Plow, to help keep their insurgent activities in check.

When bamboo was encountered, other vegetation was not well represented. Somewhat insidious, Bamboo tended to take over some areas and choke other plants out of existence in the process. At some point, I had noticed that the soil composition within the more prolific areas of bamboo was quite different than in other vegetated areas. It appeared to be lacking in humus and other essential ingredients that otherwise help to maintain a rich, fertile soil. However, in the absence of well-composted topsoil, bamboo still appeared to thrive in its meager sandy clay-like, gray sub-surface, without needing those rich earthy nutrients to flourish. In woodland jungle, on the other hand, everything seemed to grow so well there also. With the help of a naturally composted soil that perpetually renewed itself within these high-canopied tropical forests, its hardy, robust plant-life was complimented and quenched further by the region's constant humidity factor, and the relentless season of Monsoon.

An inadvertent positive benefit that Land Clearing had provided for the Vietnamese as a bi-product, in its efforts to help curb enemy activity, was the later realization of certain economic growth. As a result of these efforts, many large tracts of cleared land had later been opened up for farming and building construction, sometime after the war had wound down and finally come to its conclusion.

My plow, #26, shrouded by the thick bamboo

In more recent times, environmentalists had brazenly accused us (the U.S.) of raping the jungles of Vietnam, thereby destroying their rich tropical soil composition, as we purportedly left vast open areas of wasteland in our wake. They further claimed that, through the use of toxic Agent Orange chemical compounds, 2-4-D and 2-4-5-T, along with the even more immediately effective suppression of plant life, which involved the deployment of our Rome Plow tractors, that it all dealt a devastating blow to the overall environment and natural habitats within the region.

Regarding the overall use of Agent Orange, they may actually have a valid point, as I considered it to have been a huge mistake on our part to employ such chemical agents over there. I don't believe we thoroughly knew just what that stuff was, inasmuch as we may have known about what it would ultimately do to vegetation. Actively effective in its systemic process, contaminates from it also found their way into the soil, and consequently rendered those areas useless for farming, or for much of anything else for quite some time. Thereafter, when the Vietnamese or native Montagnard peoples came in contact with this chemical or its residue, as it would eventually find its way into streams where water was regularly drawn and ingested, some of their expectant mothers consequently gave birth to children exhibiting pronounced physical defects that were directly attributed to the presence of this toxic agent. Over the years, some of our own military people had also paid the ultimate price for the use and distribution of these defoliants. While many of those who had been directly involved in the handling process were largely unaware of this chemical's degree of toxicity, they eventually succumbed to its deadly after-effects which were often left upon the body.

But, the environmentalists were largely wrong about the cleared areas that our mechanized Land Clearing operations produced. In the interest of preventing erosion the severed shoots of the plants we cut were left standing, for the most part, to continue to regenerate themselves over the post-war years. That way, their subsequent re-growth would eventually restore many of these cleared areas to their former prolific state within a certain period of time. In some of the other areas, the Vietnamese, in later reclaiming the land, removed those remaining root structures to alternately use the acreage for farming, and/or new home construction. Conversely, it would have taken them years to remove all of the dense jungle foliage by hand, in order to fully utilize the land in this way. In effect, we had saved these people the trouble of having to clear the land for themselves, which, in turn, fully suited their ongoing building and farming interests. However, this wasn't actually brokered in as part of our objective, as our only real focus was on improving security throughout the region.

In the Xuan Loc area (*pronounced, Swan Lock*), which was nearer the coast and roughly 95 kilometers (*60 miles*) northeast of Bearcat, we picked up our security, a mechanized company from the 11th Armored Cavalry Regiment (*out of Blackhorse base camp*), and found woodland-type jungle most of the way along QL-1 highway, with some flat and rolling hill type terrain encountered. Oddly, the security was getting a little frustrated at the time, after establishing part of their outward defense system. They were setting up booby-traps in the area on their Claymore mines to target any enemy elements that may have been intent on trying to breach our encampment.

Claymores were a type of land mine that were ordinarily set up on the ground's surface, with single or multiple trip-wires laid out in advance of them, linked to a triggering mechanism. This particular mine, which had a slightly curved front and was filled with buckshot, was placed at an angle on the ground, to spread its spray of lead shot in an upward, fanned out trajectory, when tripped. Carefully covered with light branches or leaves, it became effectively camouflaged, where it remained largely undetectable to most humans and animals alike.

The source of their frustration was that, in the night, the Viet Cong were actually booby trapping the previously set booby-trapped Claymores, effectively toying with our security and showing them up. Fortunately, this 11th Cav mechanized unit detected it without further incident and quit using them there as a result of the little 'tit-for-tat' game that was being played.

Taking a short smoke break along QL-1 highway, near Xuan Loc, with a shaded grove of rubber trees in the background. This particular area was just a small part of the Michelin Rubber Plantation.

At some point along the way, we happened to run right up on a shady grove of rubber trees that happened to lie within the sprawling Michelin Rubber Plantation, which was owned and operated by the French. It was one section of a large, commercial rubber tree orchard that had featured their organized stands of countless trees, which stretched out over miles in some directions. We had not encountered these since operation Gainesville, where we earlier camped among similar tree groves in the Lai Khe area, north of Saigon, when working for the 1st Infantry Division ('Big Red One'). It was nicely shaded and cooler there in these rubber groves, which had always offered a welcome change from the usual glaring, hot, sticky conditions.

I remember that we had come to a stopping point upon entering this plantation near Xuan Loc, and had to wait while our brass conferred with these Frenchmen about cutting down their rubber trees that were directly in the 'cut line', along one side of QL-1. Finally, the word was given and we off-loaded an anchor chain (*just like those on ships*) from one of our trucks and attached each end-link, with a draw bar pin, to the rear tow receivers of two plows. These two plows then aligned themselves in the cut, with one tractor roughly 6 to 8 rows deep, within the grove of rubber trees, while the other rode just outside the woodline. They moved along in unison, set abreast of each other, with the big chain dragging along behind in 'U' fashion, effectively pulling down all the targeted rows of these shallow-rooted trees as they went. Behind them, a bull blade dozer pushed all of the downed trees into piles to be burned later. This would be done until enough of these trees were cut back to match where the woodland jungle cut line had left off and began again, roughly about a half-mile or so down the road. It was initially met with protest from these Frenchmen, as I recall, until a settlement of sorts was eventually reached by way of radio through our higher command, (MACV) in Saigon.

Essentially, our tax dollars later reimbursed them for the loss of these trees, and compensated them additionally for what the trees would have otherwise produced over their expected life span.

On a much later outing, northwest of Saigon, near Song Be (*pronounced, Song Bay*), we encountered groves of fruit trees, which produced very large, melon-size fruits that were ball-shaped and had a hard rind with spiky protuberances, the likes I had never encountered before. They hung plainly from each of the tree's branches like large ornamental balls on a Christmas tree, although, these trees appeared somewhat similar in structure to very large apple trees. It had been suggested that they were probably Breadfruit, but some later inquiries revealed them as Durian Fruit, a strangely exotic fruit that's native to Southeast Asia and emits a somewhat foul odor when ripe. It is quite popular and considered to be an expensive, yet sweet delectable treat to natives of the region. However, when this fruit is fully ripe and being consumed, the smell is so disgusting, it is said that it's like dining inside a public restroom. With that unappetizing description in mind, it's hard to understand why this odd sort of fruit would be such a delicacy, although the Vietnamese people tended to acquire some awfully strange tastes in exotic foods.

Durian Fruit

Unfortunately, from our position, these unusual fruit-laden trees also lay within the appointed path of the existing trace line, and had to be uprooted and removed before continuing on, just like the rubber trees that were encountered. So, once again, we waited there for several minutes with our engines idling while our brass conferred with the higher command in Saigon for some kind of formal decision, one way or another.

The old Vietnamese farmer who owned the trees pleaded with our people to rescind the order as we were told to proceed. Upon hearing the final word, he broke down and began to cry, as he scrambled about in a weak attempt to wave off the plows. Sadly, we were destroying much of his livelihood that kept his family clothed and fed; trees that he likely nurtured to maturity and harvested every year. But, it was not something we could change, although it was a tragedy of sorts for him. It was all part of the necessary clearing process, to discourage the easy ambush tactics that enemy elements would persistently employ. There was just no room for any compromise in these instances.

In this case, it seemed unlikely that the fruit tree farmer ever got compensated for his loss, as compared with the French.

~ Nine ~

A Veritable Zoo

One of the constant annoying hazards that all Rome Plow operators were plagued with from time to time was a belly pan fire, which would commonly occur at least once during the course of a week's activity out within the cut. Belly pans were heavy gauge steel covers that were bolted beneath the plow's engine and transmission to offer further shielding for those two vital components. Because of potential land mines and harsh terrain, these shroud-like steel pans became necessary protective additions to the tractor's otherwise vulnerable undercarriage. Given that they were vented to dissipate the heat, the only problem with them was that vegetation tended to find its way into these thick steel shrouds, entering through the vent holes to effectively block any beneficial air-flow. As a result, the un-vented heat would build up within the clogged pans to the point of eventually igniting the accumulated dried material. The operator would then have to stop and try to put it out. The difficulty with this was that the fire was way underneath the plow, and the only way to properly address it required crawling under the body of the tractor to clumsily attend to it, while partially lying down on the ground, quite often with the engine still running.

To further compound the problem, there were no fire extinguishers aboard at that time; we rarely had them. But, we generally kept a 5-gallon water can within the cab that was normally reserved for our radiators, and also for fighting these bothersome little 'nuisance fires'. The only thing was, it took a little more time and effort to transfer water from the 5-gallon 'Jerry can' into a small pail, or into one's steel pot (*helmet*), whereby it could be splashed in through the vent holes, although awkwardly done, to effectively extinguish the smoldering fiery threat. In the absence of ample water for battling these fires, scooping loose soil up into cupped hands and tossing it in through the vents also turned out to be a fairly useful way to get it out. This practice simply smothered the debris inside to where it could no longer burn, although it then tended to block the vented air flow until such time when the belly pan could be dropped down and cleaned out later. Additionally, banging one's head on the steel bottom of the plow was something that often 'went with the territory', when addressing one of these occasional, annoying little fires.

Out of sheer determination, sometimes one had to be as diligent and resourceful as possible in resolving these sudden emergencies, as it was mostly left to each operator to extinguish their own flare-up. Quite often, there just wasn't anyone else around to lend a helping hand whenever these fires occurred. In the remote event where he just wasn't able to put the fire out, the flames would usually flare up further, within the engine compartment and burn up component parts and essential wiring, ending the day, if not the week, for that particular plow.

In response to these ongoing difficulties, we later acquired some 5-gallon sprayer cans, complete with a hand pump and hose for an easier time of it.

After noticing a fair amount of light-gray smoke rising from below on one such occasion, I stopped my plow and crawled under; managing to get my fire out fairly quick by smothering it with loose soil, before discovering that my blade had sheared off the top of a giant termite mound. Pausing to look around, I suddenly found myself lying on top of its exposed inner labyrinth-like earthen structure, with a steady flow of seemingly agitated and disturbed termites emerging all around me. Not surprisingly, I just happened to bang my head while hastily getting out this time, with a mess of these giant termites crawling all over me. It was downright freaky. While unusually large for this type of insect, at about an inch in length, they had these big scary looking mandibles that broke one's skin when pinched (or bitten).

Reacting desperately to this sudden chaotic situation, I really didn't care so much about my head hurting from the contact of skin and bone on steel as much as I urgently needed to get out from under there and shake all of those pinching bugs off of me. Exiting from below, my frantic reaction must have looked like some sort of weird animated moment, as I hurriedly scrambled out and got clear of the tractor. I then nervously jumped up and down, while flailing my arms about in a desperate attempt to knock them from my head, chest, and lower extremities, all the while groaning in disgust and yelping at each painful pinch. Having neglected to properly blouse my fatigue pant-legs, I found their presence to be so complete that I even had to drop my pants and frantically pick a few of them out of private areas, before finally regaining my composure and resuming further activity around the trace.

Bees had also posed a particularly distressing problem for us from time to time, as they tended to give us some serious reason for concern. While continuing to slice our way through the heavily foliated jungle, our activity occasionally stirred up a nest of them and some of us consequently got stung while trying to flee from their angry swarm. As they reacted aggressively to our destructive presence, the plow's screened cab helped a great deal in deterring many of them from coming directly in on us; although, some would soon manage to find their way around and get in anyway, through the open front of the cab, to vengefully hit their mark.

In response to this anxiety-producing situation, setting off a green smoke grenade within the cab was found to be the most effective way to quell the problem and keep them out. For reasons unknown, green smoke always seemed to work the best on bees, while other colors had only a marginal effect on them. To escape the furious wrath of a stirred-up nest of bees, an operator without the aid of green smoke was optionally compelled to just put the tractor in 3rd gear and simply hightail it out of their frenzied, chaotic domain, essentially exiting the immediate area in full retreat. But, when green smoke happened to be at hand and was 'popped' within the cab, the unpleasant smell of sulphur from it was found to be somewhat uncomfortable and a little nauseating, as I would typically struggle to breathe through a rag. Although having been somewhat difficult to endure under the extreme circumstances of this otherwise unsolvable problem, the unusual use of smoke in this way ultimately proved to be quite a worthwhile trade-off, and a welcome solution in the end.

Animals and wildlife in general, in and around the jungle, were often quite fantastic and unbelievably different than most that I had encountered prior to Vietnam. The Mouse Deer (*Tragulus Napu*) was a case in point. It stood only about 18 to 24 inches high, with long thin little legs that had what appeared to be tiny cloven hoofs; however, upon closer inspection, four little digits were actually revealed on each narrow, petite-sized foot. With the basic form of its diminutive frame looking so very much like a miniature deer, its head more closely resembled that of a rodent. Being related to neither a deer nor a mouse, they were found to be docile little creatures, which were actually quite cute and were supposedly more widely identified and recognized as Vietnam's national animal. While fairly easy to capture, they often weren't taken for their meat as much as for their adaptability in becoming tamed backyard pets. Unlike actual deer, they didn't sport the usual display of skull horns, though the males instead featured small upturned pointed tusks. These protruded from within their lower jaws, much like the tusks that ordinarily protrude from the jaws of male wild boars. Interestingly, the French, in their earlier occupation of Vietnam, had commonly referred to these unusual creatures as *Chevrotains*, which was taken from the word *Chevre*, meaning goat.

Two examples of Malay Mouse Deer, or Chevrotain.

In Bearcat, Praying Mantises were often seen clinging to the ceilings on the porches of many of the screened tent buildings there. Although these were very large ones, at about six to eight inches long, they were totally harmless despite their extraordinary size, and were actually quite beneficial to our presence there since they helped to thin out the varying amounts of other pesky insects in the vicinity.

Snakes were also encountered and avoided in certain areas, as we continued to work our way through the jungle. Among those noted on occasion, were various Vipers, Cobras, and a few large Pythons. On a rare occurrence, I frightfully encountered a Bamboo Viper that somehow slipped in through the side screen of my plow's cab, near the hydraulic tank, while I was cutting in thick bamboo one day. This little snake is only about a foot long, its body about a half of an inch thick and colored a distinctive light green to blend in with its suited bamboo habitat. Fortunately, it just happened to catch my eye when it suddenly came slithering in through the screen. At the time, I remembered some earlier hearsay about this tiny notorious snake, which suggested that while it was so lethal, it was loosely referred to as, a *'Step 'n Half.'*

Legend has it, (*so I was told*) that after being bitten by this particular snake, one could only go a step and a half before immediately giving up the ghost...thus the name. Somehow, while observing this little serpent, I wasn't quite willing to doubt the legend and put it to a test. My heart quickly began to race as I anxiously watched it slither along the outer edges of my plow's rectangular-shaped hydraulic tank, just inside the screen. I then noticed that the reason it purposely kept to that very outer edge was due to the other surface areas being much too warm to traverse, from the hot hydraulic fluid within. Keeping my eyes trained on this little unwanted reptile, I slowly stood up and took up my rifle from behind the seat, allowing the plow to continue on without me. I then used the butt of it to persuade the snake to turn back and go out again through the heavy steel-meshed screen. As it finally went back through and quickly dropped down onto the track below, I breathed a huge sigh of relief and immediately refocused on my primary task at hand. Sitting back down in the operator's seat, I found that the tractor had only strayed slightly from its briefly abandoned course, within the area's seemingly endless sea of tall bamboo stalks.

While regularly positioned out around the established trace, our assigned mechanized security force once came across an exceptionally large Python that was desperately trying to escape our pressing demolition of its established habitat. As its capture quickly unfolded before me, I watched about eight of their guys holding it up at intervals, while another took the celebrated snapshot before they finally released the monster snake back toward its previous outward course of frenzied flight.

Once, while steadily moving along in the cut, I spotted a mother Orangutan clutching its newborn and fleeing for its life, as it escaped the imminent onslaught of the plows. Quickly ambling across the debris-laden expanse, this large, reddish-brown primate soon vanished nearly as fast as it had appeared, taking immediate refuge within another nearby, unmolested woodline.

On yet another outing, I caught sight of an unbelievably large, pinkish-gray colored monitor lizard, just about 15 to 20 feet out from the left side of my plow. It was frantically fleeing from the ground-shaking chaos that our machinery was causing within its formerly tranquil, wooded sanctuary. While uneventfully moving my tractor along in the cut, I steadily worked my blade within an area of small trees and light shrubbery. Then, I noticed this large lizard suddenly zip by me as it hurriedly scrambled along on its belly and raced ahead; slipping through a thicket of small trees with its four muscular legs churning up the soil in desperation as it went. In this sudden surprising encounter, I could only sit there in amazement and marvel at it. It was, without a doubt, a most impressive lizard. Having been the largest reptile of this type that I had ever seen up close, I found that, aside from its unusual coloration, it amazingly resembled the famed Komodo dragon, which I had earlier seen photos of. Judging from its close proximity to my plow, I would say that it was easily about 6 to 7 feet overall in length.

In an altogether different encounter, one of the mechanics riding in our M-548 maintenance track, who had somehow appropriated a WWII version of a Thompson sub-machine gun, suddenly decided to put the weapon to a more resourceful use. At some point during the day, he happened to spot a wild boar quickly exiting the woodline, while they were momentarily parked out in the cleared area surrounding the cut. As the animal desperately raced across the clearing before them, he gave it just a short burst of about three or four rounds, which immediately brought the hairy beast down. A few of the others aboard helped in gutting and dressing the animal out, after which they eventually tied it off at the rear of the maintenance track, allowing for it to just hang over the side to bleed out.

Wild Boar, hanging dressed out prior to our projected barbecue.

While cleaning the carcass, the liver was retained and one of our officers sent a small portion of it back on an outbound chopper to have it tested. With this boar carcass hanging over the side of our maintenance track, we all entertained the sudden prospect of having wild game on our plate, as many of us anticipated a barbecue that night. As it turned out, the liver test revealed that the animal had worms; so, instead of keeping it, our guys just let the nearby villagers have it. They accepted it with glee, despite the latest findings. A few of the local village men seemed to acknowledge the meat's condition, and insisted that it was ok, indicating that they could simply cook it out, which I guess they could. However, our brass simply didn't want to chance anything with the tainted meat, on our behalf.

Flying Lemurs (*also known as Colugos*) were regular occupants of the jungle's large tree canopies and were often seen soaring from tree to tree, as they noisily moved about. They were noticeably upset by our mechanized activities below, which were disturbing their typically tranquil setting high above the forest floor, where they seemed to suggest that they were the lords of that lofty domain. Colored light brown with white splotches, they looked like a cross between a monkey and a big Bat. With unusually large bug eyes and a fair amount of flabby excess furry skin, they frequently stretched both sides of it out wide to create an airfoil for gliding, much in the same manner that the more common flying squirrel does.

Cutting down the large trees that they inhabited usually brought about some commotion and noisy protest from these treetop tenants. Their frantic and furious presence was clearly noticeable, although mostly drowned out by the even noisier ruckus of the plows. While their protest was briefly noted, it was soon rebuffed by the mechanical swordplay that ultimately dropped the trees to their new horizontal position down on the floor of the jungle. In response to this, these monkey-like furry flyers quickly glided off into other adjoining tree canopies to continue their feverish demonstrations of dissent.

In the interest of making sure snipers couldn't use these large trees to pick individuals off as convoys passed by, the more experienced operators would stab their 'stingers' into the tree trunks, taking chunk-size slices of wood away in the process, until they could eventually get the tree to crack and topple over. The important thing with this practice was to make sure that it was done in such a way as to fall cleanly, away from the plow, and away from anything else that might have been in harm's way. This was a practice that, generally, only the more skilled operators were entrusted with (*As time went on, that standing rule eroded somewhat, and inexperienced operators tried their hand at it as well*).

Otherwise, the trees were usually left for others to detonate later with C-4 plastic explosives (*which offered a distinctively different way of bringing down large trees*). Many of these appeared to be large Teak trees that featured tall, thick trunks along with their high leafy canopies. If one happened to come errantly crashing down onto a plow, it would surely crush the reinforced steel cab, leaving the operator either killed, injured, or trapped within.

The extended wedge, or 'Stinger', featured on the angled K/G blade, was a very efficient tool, which was mainly used for slicing into the trunks of larger trees in order to more effectively bring them down. The angled blade edge and brush guards were used more exclusively for cutting lighter trees and shrubbery, as this blade couldn't effectively penetrate most of the larger-sized trees, like the sharpened Stinger had been designed to accomplish.

On one particularly notable day, while working an operation out within the Iron Triangle area near the Cambodian border, I was busy doing some repair work on my plow inside our NDP, when a serious mishap suddenly occurred. Oddly, I just happened to look up from what I was doing and notice a sizeable tree come crashing down on one of our other plows as it worked the cut nearby, just a short distance outside our perimeter berm. As the tall tree impacted on the tractor, it somehow crushed the cab in such a way that the operator's head was left stuck between the crumpled cab and the top of the fuel tank (*with him still partially seated in the operator's chair*), and he just couldn't get it out.

He was alright, though, other than suffering some minor cuts and contusions after becoming stuck there in the cab's steel encased wreckage. Several attempts were made to extract him from it, but all efforts at that point were to no avail. Offering to help, our security brought in an M88 Tank Retriever to assist in lifting the heavy tree off of the plow's critically damaged cab. It then hooked its boom to it and lifted in unison with the aid of two other plows to raise the tree from its cradle on the seriously indented and crushed steel cab. At that point, several other men with 60-inch pry bars then managed to raise the mangled cab just enough, from its floorboard mooring, to get the man's head out.

A few hours after that freak incident occurred, a platoon of our mechanized security returned to our NDP and were suddenly engaged in a firefight with several Viet Cong guerrillas a short distance away. They sustained a casualty sometime during the skirmish, who was immediately rushed into our compound for emergency medical assistance. He was quickly brought in on a stretcher and gently laid onto the ground within our area, as his APC then headed back out into the firefight. Immediately, our medic began working on him and, at some point, inserted an IV tube. Taking up the IV bottle, he glanced over in my direction and instructed me to hold it up for him while he continued to assist this unconscious, wounded man. It was somewhat of a bloody mess to look at, but the medic soon managed to stop the bleeding in the soldier's lower abdomen, and shortly after, got the man 'dusted off' to an Army field hospital. As I stood there holding the IV bottle and averting my eyes from the not-so-pretty scene below, one of our Lieutenants suddenly rushed over and took it from me, quickly giving me instructions to go join the other operators and help secure the perimeter. Most of our security, in responding to the firefight, were now entirely outside the berm. In this situation, it was then quite possible that a few of the insurgent VC might manage to get by them and come on in. Fortunately, that didn't occur, but we had to consider these things when the action was taking place just outside of our Night Defensive Position.

We had always loosely referred to the enemy as 'Charlie', which was the NATO phonetic alphabet's term for the letter 'C', in referring to Cong. More specifically, VC was used for Viet Cong (Victor Charlie). The North Vietnamese Army regulars, who were occasionally captured or killed by our security, were simply referred to as, 'NVA'.

When working in predominantly thick woodland-type jungle, 'Windrowing' was a common practical measure that was sometimes called for and put to use. In application, downed trees and large shrubbery were systematically pushed into piles that ran in a line centered within the clearing, between the newly established woodline and the roadway, for purposes of burning it off. This practice ideally left the cleared areas entirely free of debris and downed cover, while helping to discourage any potential ambushers who otherwise might have considered using it for their hit-and-run style of warfare.

On one of my not-so-pleasant workdays, I was assigned to the somewhat rare task of Windrowing. Oddly, just as I was rounding up some of the scattered downed debris that generously filled the clearing, a sizeable branch errantly wound its way up and into my cab, which I mindfully kept in view out of the corner of my right eye. Sitting there watching it wriggle and twist its way in, as the plow continued to move along with the blade's accumulated mass of broken limbs and the like, I mistakenly assumed that it would probably wind its way back out again like other invasive tree branches in the past had always done. At that point, I paused and wondered if I should just stop, get out, and clear it by hand. Instead, I opted to keep going, thinking that it would once again clear itself, much like the other wayward branches had done before it. But this time, it came all the way in and quickly lodged against the back screen of the cab, where it began to bend like an archer's bow. Then, just as I made a slight turn, it quickly let loose with a spring-like force that smacked it hard against my right cheekbone. Consequently, it had knocked me unconscious for a short period of time. When I came to, somehow still halfway seated within the operator's seat, I found that the plow was still steadily moving along, and noticed that I had traveled nearly two hundred yards down range. I then stopped the tractor and got out, feeling quite dazed and somewhat disoriented. Suddenly, the right side of my face began hurting from the blow, and I felt a serious headache developing. Indeed, that branch had eventually wound its way back out, as I had earlier surmised, for it was nowhere to be found after having its way with me. Shortly afterward, our Deuzenhalf came on the scene to investigate my reason for stopping and I was soon sent back to our NDP with a lacerated cheek that took a few stitches to close, along with a splitting headache that turned out to be from a mild concussion. Someone else on the truck then took over my plow for the remainder of the day, as I was definitely done.

Windrowing downed debris within heavy, woodland type jungle. This shot was taken with my old Polaroid camera, somewhere in the Xuan Loc area, off QL-1 highway, east of Bien Hoa. The time was roughly around June or July, of 1968, when we worked with the 11th Armored Cav Regiment out of Blackhorse.

Although I was very fortunate to escape serious injury and didn't receive any enemy-induced wounds while operating a Rome Plow like many others did, I certainly had my fair share of cuts, abrasions, burns and contusions, along with a few stitches here and there. The almost daily occurrences of suffering minor injuries along the way, was entirely normal and even somewhat expected for the kind of work we took on. Quite simply, our nagging little injuries, while not being entirely problematic, tended to go right along 'with the territory' of our Land Clearing missions, as most of us just sucked it up and resumed with our assigned duties. Regarding enemy action, needless to say, I wasn't exactly despondent over my lack of Purple Heart medals. Fortunately, throughout my entire time with LCT, I found that I happened to be just a bit luckier than some others were out around the trace.

~ Ten ~

Stand Downs and R&R

When each 45 – 60 day field operation finally came to its point of completion, our 10-ton tractor/trailer rigs (*lowboy*s) would soon reappear at our final NDP site, where we'd pack everything back up again for the return trip to Bearcat. With our mission completed, it was then time to head toward home base once again, for our regularly scheduled 15-day 'Stand Down'. These periodic Stand Downs represented a timely renewal process for our tractors and other equipment, as well as for the field-weary men of the unit. They allowed us some time to clean everything up, re-tool, and give everyone a needed break from life in the field.

Back at Bearcat, we had hot showers again, beer, movies in our outdoor theater, and clean clothes every day. These were indeed welcome changes from roughing it out in the 'boonies', and were simple pleasures that we hadn't appreciated quite so much prior to going out on our rather lengthy field operations. At that time, haircuts suddenly became a top priority, along with a proper shave. After 45 to 60 days of 'camping out' in the jungle, our brass and NCOs saw to it that we should finally get our outward appearances fully restored to look regulation-normal again. At night, the NCO club and the Enlisted men's club gave us all a chance to catch up on some of those beers and other adult beverages that we had missed, along with the added attraction of live music, performed by a few of the marginally talented young local Vietnamese men and women there. They often played former rock 'n roll hits, and were seen as just ok, musically. But, their Vietnamese-accented form of English tended to butcher the songs' lyrics to the point where nobody really cared much for their efforts. They tried to imitate popular artist's original presentations while they played, but it tended to just pass as background music. Many of the guys just wanted to talk and joke while they drank anyway, since we didn't have any girls around there to dance with. *This is where the Navy usually had it over us, with their occasional visits to different exotic ports of call, where women were generally always available.*

Optionally, we could always purchase canned U.S. domestic beer and sodas from the nearby PX to keep readily on hand in the barracks. While some of us didn't care to blow our money on the EM club, we otherwise kept small ice chests in reserve for our own personal use. Block ice was the only type of ice available to us there, which actually lasted somewhat longer if mainly kept in its full rectangular form. Using it that way, we learned how to effectively chill our canned beverages in relatively short order. Whenever a beer or soda was desired, one could simply roll their own canned drink one at a time, by continually spinning the can with palm and fingers as it lay on the ice block, until an impression from the can's ridges formed in the ice to help keep it positioned there. At that point, it could then be spun at high speed with one's fingertips until it was soon quite cold. Amazingly, it only took a few minutes to skillfully "roll your own" can of beer or soda to a decent chill. It proved to be well worth the effort, without having to chop up the bulk of one's ice to chill all of their canned drinks at once. In those days, most of the cans had ring-top pull tabs on the tops of each can for accessing the carbonated liquid within. That style lasted until the current StaTab type opener was later introduced. Also at that time, there were still some cans around that required a 'church key' with which to puncture a couple of 'V' type holes on the tops of each one for pouring and venting, prior to consuming its contents.

Since we didn't have any TV there, the outdoor theater made up for it by showing some fairly good 'flicks', out within its own designated area. We often immersed ourselves there in the comfortably warm, relaxing night air, which frequently accompanied a clear, starry sky overhead. When one of the films happened to feature a sexy actress, whistles and catcalls would always come out, as we tended to enjoy these evening diversions from the war. Aside from that, on occasion our brass would mysteriously procure an excess supply of steaks somewhere (*nobody asked*), and we would all gather for a company barbecue or cookout, which became a nice little morale booster for us. Otherwise, maintenance was the order of the day, as we continued to faithfully prepare for our next operation. With numerous acres and hectors yet to be cleared, there was naturally always another field operation to come.

We all enjoyed our occasional Company barbecues in Bearcat.

During our maintenance Stand Down, the plows and bull blade tractors were usually lined up abreast of each other in the motorpool area, where we would disassemble component parts for repair or replacement, as determined by our maintenance people. Since there was sometimes a short supply of certain replacement parts or components, there was occasionally a 'dead plow' around (*from being irreparably damaged in enemy action*), which could be cannibalized for the sake of the others when needed.

At this time, our maintenance welders would repair damaged cabs, brush guards, blades and anything else that needed to be re-secured or reinforced on our tractors. We would also take this opportunity to pull the big radiators out and flush the cooling system. It also sometimes became necessary to replace any of the more seriously damaged radiator cores with new ones. Engines and transmissions would also be checked out thoroughly and were re-tooled as needed. Essentially, all of the plows were given a top to bottom physical at this time, and were repaired or maintained accordingly toward a clean bill of health, as we thoroughly prepared for our next challenge in the field.

One of our maintenance crew members making some welding repairs.

Whether we were out working in the field or tooling our equipment during our periodic Stand Downs, we had always worked closely with our maintenance mechanics, which proved to be quite beneficial to all of the operators. It helped us to become a little more knowledgeable about the machinery's component parts, and to generally get a better overview of a D7E's operational limitations. Getting oil and grease all over us in the process was just a natural part of things when repairing and replacing certain component parts. Some of the more specific parts that were slated for replacement often required a degree of strength and tenacity, as well as knowledge, in getting them set correctly into place before securing with screws, bolts or pins. In that regard, expletives frequently filled the air space around the motorpool. Our doggedly determined efforts sometimes resulted in busting our knuckles or banging our shins somewhere out upon the tractor's steel surfaces. As a result, we were occasionally compelled to perform our own little pain dance for a few brief excruciating moments.

For those of us who may have operated our plows somewhat recklessly in our earlier months there, we soon learned the overall value of having a tractor that performed consistently well for us. Over the course of time, most of us naturally adjusted our ways to gain a more respectful regard for our machinery. While the change in attitude tended to create a little more detailed maintenance, it also gave us a more heightened concern for our assigned piece of equipment. In turn, this translated into fewer breakdowns in the cut, with more enemy cover being removed from the particular targeted areas of thick vegetation.

Here, I was working on my first assigned tractor, # 22, prior to adding the KG blade and tree cab. Below, a squad leader supervises work on our radiators, during Stand Down, as another operator ('Cornbread' Jones) thoroughly cleans his plow's giant radiator core.

It was a conscientiously cooperative sort of relationship that we tended to maintain with our crew of mechanics, given that both operator and mechanic alike would generally benefit from the mutual support. Our helpful efforts often resulted in less of a workload for them, and vice versa. However, some of us would have much rather preferred the usual activity of crawling our plows around the cut than to be left spending any 'dead time' helping them with major repairs around camp during the day. Whenever we suffered a breakdown of any consequence, it was always expected that we stay with our piece of equipment throughout the down time in order to assist our mechanics in any way. Our joint effort usually served to get the machine back on line quicker, where we could resume with our normal daily activity, in working the cut. The only exceptions involved the few impaired plows and bull blade tractors, which had suffered blown engines or transmissions along with those others that had been severely damaged by land mines or other such enemy generated action. In those instances, operators were typically re-assigned to another plow. Depending on the severity of damage or particular part replacement, the downed one was usually sent back to our main base camp for major repairs. If there were no extra tractors immediately available, the operator simply had to sit it out for the most part, until one came available again. Whenever a tractor became a total loss because of enemy activity, and could no longer be fully restored, it was commonly set aside for cannibalization, while a new replacement tractor was usually requisitioned from the states to take its place. Typically, it could be received within about 3 to 5 weeks, and was then subsequently outfitted with a cab and K/G blade before sending it out to mingle with the rest of the herd.

22, ready for repairs and new component parts. Here, it stood in the motorpool area in Bearcat with then Pfc. T. Brown on the left, and Pfc. R.L. Tluchak on the right.

In addition to the usual periodic maintenance and certain necessary repairs, some touch-up or full coverage painting of the plows was called for during these Stand Downs. At that stage of things, industrial grade 'olive drab' enamel was generously applied to its surfaces, keeping with the typical uniform color of everything else that naturally smacked of Army green.

As I happened to be working in the motor pool one day on one of our 'Stand Downs' in Bearcat, I noticed that it was just shy of noon and started for the mess hall with a few of the other men. It was clear across the company area, which ordinarily took only about 5 or 6 minutes to traverse. We arrived there a little early, so the line was building, with the place due to open up in just a few more minutes for noon chow. While standing there fully engaged in conversation, suddenly, a shot rang out, which came from a distance of about 80 yards behind us. Immediately, several of us dashed over to where Headquarters Company and A Company came together, to see just what might have actually occurred there. Approaching the apparent scene, we then noticed a lightly complected black man down on the ground with others over him. A couple of Medics who had quickly arrived on the scene tried desperately to revive him despite the bullet hole that all could see on his forehead. As we stood there watching, from just a short distance away, we knew we were pretty much looking at a dead man. Before our eyes, he turned several shades of purple before a Huey Med-Evac chopper finally descended onto the scene, stirring up a lot of wind and dust as it landed quite near us. Quickly loading him onto a stretcher, they safely secured it aboard one of the chopper's skids and signaled the pilot to take off. The chopper then immediately lifted off and disappeared out of sight. After holding onto our hats while waiting for the chopper's dust cloud to clear, we could only look at each other in shock and amazement, from the stunning sight that had just been placed before us. Slowly, we then turned and sauntered on back to the mess hall for lunch. How we could manage to eat after witnessing that, I couldn't begin to explain.

As it turned out, there was a personal feud of sorts going on between this man and a young Hispanic within the nearby barracks in A Company. Soon after, the shooter ran to his Orderly Room and turned himself in to the 1st Sergeant there. He was later charged with murder and sent to the infamous 'Long Binh Jail,' otherwise known as LBJ. I imagine he may likely be in Leavenworth Penitentiary today.

In an unusual turn of events, just as we ended one of our periodic Stand Downs, and were then about to depart on another, I was chosen to instead, load up my plow and travel a couple of hours northeast; up QL-15 and out over QL-1 highway. According to my platoon sergeant's orders, I was to report to Blackhorse base camp, home of the 11th Armored Cav Regiment for some solo tactical cutting. It was required that I work there, TDY (*temporary duty*), for about 10 days, because they needed to increase their outside perimeter area to enlarge the camp. They also wanted me to cut out a new LZ for them, to better facilitate the occasional arrival and departure of helicopters. So, as it came about, the lowboy driver turned his rig around after I offloaded my plow there, and he headed on back to Bearcat while I remained in Blackhorse for the duration of this designated mini-operation. However, it actually turned out to be a real 'gravy job' for me. They took special care of my needs and provided me with more security than usual, along with an air compressor and grinder, and maintenance people to call on whenever needed.

Throughout my time there, I worked my plow in heavy woodland type jungle, just outside their compound, with perfectly level terrain in view. Amazingly, I never encountered any unusual hazards, land mines, or other enemy type activity while accomplishing this temporary assigned task. For that matter, I didn't even know why I was even chosen for that job. But, I always felt gratified by it, knowing that they surely wouldn't have sent a questionable or inexperienced operator out there to work it and see it through.

Soon after that, I was promoted to SP/4 (*Spec-4*), which is an E-4 specialist rank, equal in grade to the rank of corporal. Recognized in its 'subdued' form, the designated sleeve patch featured the symbol of a black spread-eagle, centered on an O.D. green shield.

During the long course of one's Vietnam tour of duty, a brief vacation of sorts usually presented itself in the form of an R&R (*Rest and Relaxation*) period. After 6 months of service in country, an opportunity to relax and unwind away from the war was commonly given. Facilitating this, the U.S. government sponsored one-week trips to exotic vacation spots for all Vietnam duty soldiers, sailors, and airmen. We had several destinations to choose from, like Singapore, Bangkok, Tokyo, Australia, Malaysia, Hawaii, and a few others. For those who may have preferred to remain in country while still enjoying some leave time, there was also an oceanfront resort area south of Saigon, in Vung Tau, which offered up peaceful surroundings and warm sandy beaches, among other things. After checking out all of the choices, I finally settled on Australia. When the time came, I took the flight south out of Tan Son Nhut Air Base (*Saigon*), headed for Sydney.

Upon arrival there, I was taken to a nice hotel in the busy downtown of the city, where my room overlooked an active thoroughfare just three levels below. Looking out my hotel window, I observed the heavy vehicle traffic, all flowing in the opposite direction from what I had otherwise been used to seeing everywhere else. The helpful travel service that handled all of the U.S service members coming into Sydney outfitted me with a decent set of borrowed civilian clothes, which were to be faithfully returned at the end of my weeklong stay.

While spending a lot of time wandering around the city, I visited quite a few shops and bought a few small items here and there. I also hungrily 'chowed down' on their classic fish 'n chips while downing a few pints of locally brewed beer in a few of their nearby pubs. I felt rather comfortable chatting with some of the nice people there, who I found to be somewhat tickled to be in present company with a 'Yank'. Much like our old Aussie mechanized security boys, these folks impressed me, as they seemed just as pleasantly mischievous and genuinely fun loving. For the most part, they were found to be a quite lively and gregarious bunch, with their somewhat distinct English cockney sounding accent. Largely bent on having a fair amount of light-hearted fun with their pub time, one couldn't help but to feel entirely entertained by a few of the more clever ones among them. Although, I must admit, I was occasionally puzzled by some of the more colorful and unusual expressions that were loosely bandied about, and often felt as if I needed an interpreter to help me understand their most unusual colloquial style and form of the English language. But, they never meant any harm to me in any way when aggressively expressing some of their friendly, good-natured wit. It simply amounted to a bit of idle bar room fun by which to pass the time, while enjoying a few good pints of their locally brewed beer.

While wandering about, I couldn't help but notice that one of their pubs up the street was set up with a separate main entry door on the left and a matching door on the right. Once inside, I noticed an interior divider panel, which separated the pub into two sections with an oval, wrap-around bar positioned in the center, between the two. It served both partitioned sides, one designated 'for men only' on one side, while the other was strictly for women. Then, as I strolled up the next block, I found another pub that openly accommodated both men and women. *Go figure.*

On the last of my days in Sydney, I walked down to Circular Quay (*pronounced, 'Circular Key'*) at the foot of the harbor, where all the ferries come in and out. From there I decided to go for a vista cruise of sorts across the bay. I had found their ferries to be in the form of more modern hydrofoils, which traveled at a much more accelerated rate of speed than I had previously known. Aboard, I met a middle-aged man who lived in the town of Manly, which is just up from the famous Bondi Beach. He just happened to be commuting home from his job in Sidney, as the hydrofoil regularly facilitated his daily travel between the two. cities. We engaged in a casual and friendly conversation and, as he realized that I was an American GI with no place to really go, he invited me to come home with him to enjoy a nice home-cooked dinner with his family. I had always thought that to be an amazingly friendly gesture from someone who really didn't know me from 'Adam'; but it was actually quite typical of the general population there. Most were found to be truly kind, caring, and good-hearted folk. A few hours prior to dining with them, he introduced me to his pretty teenaged daughter, who took me around Manly and showed me some brief sights before gathering for dinner. Later that night, I caught the ferry back across to Sydney, where I was due to board my return flight on the following day.

All in all, my R&R turned out to be a genuinely enjoyable and relaxing break from the otherwise daily grind of Land Clearing operations, as well as from the constant feelings of apprehension and tension that tended to accompany me at times when working in the kinds of stressful conditions that our Rome Plows always seemed to be right in the thick of. Australia, with its relaxed atmosphere and overly-friendly people, was, in many ways, a lot more like being back in the states. Thankfully, it gave me nearly a full week to unwind and take my mind completely off the war. In that regard, it represented "just what the doctor ordered", and it was exactly what was needed at the time. To effectively sum up my time there, the happy-go-lucky 'Aussies' had a particular signature phrase that was always so aptly put: "No worries, mate!"

From around the same time frame, I also recall that one of our mess sergeants took a scheduled 7-day leave once (*which is equivalent to R&R*), with his destination Hawaii. However, after changing into 'civvies' when he arrived in Honolulu, where he was supposed to remain, he immediately booked passage on the next commercial flight to Texas, where he and his family lived. He wound up spending his short leave time there with them, still managing to get back to Vietnam on time. He apparently got away with that, even though the states were deemed to be officially 'off limits' to any military personnel without formal written orders to that effect.

~ Eleven ~

In Convoy

In July or August of 1968, the remaining line companies of the 86th Engineer Battalion then moved out of Bearcat and traveled down into the *IV Corp* tactical zone. They arrived at their new location, near the 9th Infantry Division's re-location area of Dong Tam. Putting their construction skills to work, they rapidly set up a new Battalion compound near the Mekong Delta town of My Tho (*pronounced, Mee Toe*), creating a small, separate post, which was neatly nestled right alongside the area's namesake river. With that, LCT was left alone in Bearcat, out of range from its host battalion. All but abandoned, it seemed as though we had become our own separate entity. In some ways we actually already were, as we were then largely left to our own devices when outfitting some of the company's basic needs. While some supplies that we used to receive through the battalion became increasingly harder to get, we quite often had to settle for whatever was available to us at the time. Most items were usually acquired through our various mechanized security elements and their Divisional command, even as we worked their assorted areas of operations within the III Corp zone.

Sometime shortly after, as a result of these recent developments, LCT had completely separated from the 86th Engineer Battalion and we were indeed then on our own, at least for a short while. Their overall interest in our land clearing missions had changed, to become less important to them. Plus, the usual general assistance coming from the other line companies had then become logistically impossible, given the distance between My Tho and Bearcat. Having been newly relocated a good distance to the south of us, while continuing to sustain and maintain our needs, at some point the 86th came to the realization that it was just impractical for them to go on serving both interests at the same time. Although we continued to draw land clearing projects all over the III Corp zone, this mutual conflict of interest had eventually led to our full separation from the battalion. In effect, as this separation came about, we had then largely become an orphaned company, having to sustain our needs and fend for ourselves, for the most part. Despite the sudden change of designation, we continued to take up residency in Bearcat, working here and there, for the various infantry divisions within their III Corp spheres of influence. Our mechanized security elements then resumed their basic assistance during the interim to help supply us in times of need.

In the absence of our trusty M-14 rifles, which were turned in to the 86th supply, some M-16s, a few M-79 grenade launchers, and some old WWII era grease guns were appropriated as suitable replacements. The grease guns took a 30-round .45 cal clip and were fully automatic fire, but were really slow and would also tend to jam on occasion. They were reportedly reconditioned for Vietnam service with shoulder slings added, apparently offering improved performance from their WWII and Korea days. But, they still felt somewhat strange and were found to be considerably less reliable than our M-14s.

These old leftover relics had a sizeable bolt that slid back and forth when the gun was fired. The associated action caused the weapon to violently jerk as each .45 caliber shell was expended. As a result, this jerking motion notably lessened its overall level of accuracy, rendering the weapon much more suitable for close proximity infighting. However, as their overall reliability was often called into question, it was at least heartening to many of us that these antique machine guns actually worked most of the time, if not every time.

M3-A3 'Grease Gun'

In distributing the newly-acquired firearms, the bulk of the grease guns went to the operators, with at least the welcome benefit of being a little handier to access within the plow, given their overall short profile. Otherwise, most everything else that we needed at the time was also requisitioned and procured through our various mechanized security elements. In that regard, it was mainly in their best interest that we had whatever we needed, in order to keep the operation moving forward.

While out traveling in convoy, we were occasionally compelled to help secure ourselves whenever we rolled through some of the more notorious areas where ambushes were considerably more common and regularly anticipated. In time, we intended to cure that problem with the equally notorious presence of our plows, along with maintaining our determined stance in seeing things through in order to fully accomplish our necessary mission. Ordinarily, we weren't terribly concerned about our own security while traveling en-mass. Our assigned mechanized infantry had usually been very reliable in keeping our vehicles moving, to generally avoid incidents and ambushes when in transit. But, when heading into a few of the more infamous enemy stronghold areas, like the Hobo Woods or the Iron Triangle, security was maintained at a more heightened level. We were then required to take up our own weapons and keep a more vigilant watch over ourselves. So, accordingly, we remained overly-watchful and anxiously alert, while steadily rolling along these jungle-lined two-lane roads, which wound around and into those dreaded areas that were known to be part and parcel to the enemy's more highly infiltrated domains.

In addition to our personally-issued weapons, we had shotgun-like grenade launchers (*M-79*), and a few M-60 machine guns at the ready. We also had .50 cal machine guns mounted on swivel rings (*for 360 degree movement*) on top of our (*M-548*) maintenance track vehicles. I rode the '.50' on occasion and took in some great views of everything from up there, as we rolled through villages and over hill and dale to our next NDP site. While our single-file convoy steadily moved along, I would scan the hillsides and the roadway ahead for any unusual activity or movement. If needed, I could turn the swivel chair and gun in any direction for a full field overview, to possibly address any particular perceived threat. It was kind of like riding out in the clear open air, while sitting up on the roof of a big truck. As we moved along, I sat there clutching both vertical grips on this large, impressive machine gun with both thumbs poised over a flat, butterfly-shaped triggering device. Cocking this big gun required grabbing the sliding handle on the side with one's right hand and pulling it straight back until it locked the first round in the chamber with a click. Once the butterfly trigger was depressed, it delivered a burst of devastating firepower that could cut down most anything in its path, including fairly large tree limbs, depending on the target pattern. While I never did encounter a return fire situation while up there, I did fire it a few times when we would shoot off all of our old ammo, just as the new replacement ammo periodically arrived. We referred to these cyclical perimeter events as 'Mad Minutes', when we would all take our weapons and old ammo out to the perimeter berm in the NDP. At command, we'd then fire off what old ammo we still had left for about a minute or more, before drawing our new replacement ammo. With everyone shooting simultaneously, it was amazingly as if a large firefight had suddenly ensued. *Ammunition, in its chemical composition, generally has a relatively short shelf life for full effectiveness, and its supply needed to be replenished or refreshed from time to time.*

Supplemental to the two maintenance track vehicles and roughly 28 to 30 Lowboys with plows aboard, there were several trucks that hauled equipment parts and supplies, some of which pulled water trailers along behind. Among them was a COM truck and trailer, which housed and carried our communications equipment; a fuel truck; a few Deuzenhalf (2 ½ ton) trucks with rib-framed, canvas-covered cargo beds and optional troop seating; one or two smaller, ¾ ton 'Willys' pick-up type trucks (*also with canvas bed covers*); and a few jeeps that rounded out our rather lengthy convoys. All together, with our security mixed in, we became an even more sizeable assemblage of vehicles as we proceeded down the road to our eventual field destination.

When rolling through some of the more densely populated areas within III Corp, our progress would sometimes be slowed as the local traffic picked up and created pockets of congestion. At times, our security would be forced to run traffic blockades at intersections ahead to hold up the cross-bound traffic in some of the more congested areas. As we motored on through, we were instructed to simply lean on the horn when necessary, and not stop for anything or anyone. The Vietnamese generally knew not to venture out into the roadway when these non-stop convoys barreled past. Otherwise, they would surely get run over.

Within the larger cities and towns, the normally congested civilian traffic generally consisted of some flatbed trucks, buses, Vespa motorbikes, mopeds, and a lot of bicycles. An occasional ox-drawn cart could also be seen, slowly moving along the edge of the roadway, where it always stood out in full contrast against its 20th century motorized and pedal-powered counterparts.

Another little vehicle that dominated the roadways there at that time was a three-wheeled motorized scooter that sported a tiny pick-up truck-type body. It featured a windshield under a surrey top, and actually looked remarkably like a golf cart. It was called a Lambretta, and had a bench-type seat up front in the tiny cab, with motorcycle-type handlebars controlling the single front wheel of the scooter. Much of the country's goods were hauled to market in those tiny little Lambretta pick-ups. I even hitched a ride in one once when I returned from a seven-day leave in Hawaii. After stepping off the plane, I didn't have a ride from the airport in Saigon back down to My Tho, where I was stationed for a while in the IV Corp-delta region. So I stuck out my thumb from the side of the road and one soon produced itself. I wound up riding in the back of that slow rolling little Lambretta all the way there, keeping company with some bananas and other fresh produce that were being transported south. Upon arrival, I thanked and paid the old Vietnamese driver, when he let me off just short of the main gate. Of course, I was later told that, security-wise, it was a very dumb thing to do. I was traveling without a weapon of any kind, and with that being noted, any number of things could have possibly occurred somewhere out there along the main route.

During the course of our travels from operation to operation, we normally rolled through a range of towns and villages along the way. We occasionally passed active Vietnamese marketplaces, where people milled about to buy and sell goods, and generally interact with each other. Oftentimes, a few attractive young ladies would stroll out along the road in some of the passing cities and towns, with their lovely, shimmering long black hair flowing in the breeze. They made for a pretty picture, as they walked along and chatted with each other, while sexily clad in their quite colorful, full-length split-skirt-type form-fitted dresses, which were made entirely of silk.

Rice paddies were a common sight along many of the country's roads

We also often passed miles of rice fields as our convoy progressed, where, in some places, they covered much of the countryside with their pond-like rectangular patchwork forms. They were laid out in close proximity to meandering river tributaries, which were utilized in a natural sort of way: to flood and alternately drain these rice paddies according to how their planting and harvest cycles periodically dictated. Some of the rice farmers could even be seen diligently driving their water buffaloes along the route, while a mix of women and men in rolled up pant legs worked the shallow waters of the other paddies nearby. They all conscientiously harvested and alternately re-planted the country's 'main staple' crop, which amounted to a common variety of plain white rice. From the air, it must have looked somewhat amazing, with the square and rectangular uniform paddies creating a large aerial mosaic patchwork design upon the landscape, since it stretched the massive artwork-like form out for several miles along the road.

Workers were often seen laboring in the rice fields, as our convoy would regularly barrel past these organized parcels of earth and water.

Watching them toil as we rolled on by, it was next to impossible to distinguish between the workers to determine if any VC were actually working among them. We had heard that many of them had often led double lives, becoming sympathizers to the northern communist's embattled cause by night, while working as simple peasants by day. But, that particular detail was to be left to our security to monitor, since we were presently assigned to the duty of transporting our equipment. With that in mind,, we generally focused our immediate attention on securing ourselves and enjoying some degree of rubbernecking, given all of the various unusual, if not wondrous sights that we encountered, as we moved along in convoy.

Children selling freshwater eels, which were encased in a water-filled bamboo tube, offered up their catch at different stopping points along the road, in hopes of selling them to anyone passing by, especially curious G.I.s traveling in convoy. Typically, these eels came from beneath the shallow waters of the nearby rice paddies. The rice farmers raised them there intentionally, to mainly serve as an additional food source. Venders selling bottled Coca Cola (*which didn't taste at all like the real thing*) and Malaysian imported 'Tiger' Beer, along with Saigon brewed '33' Beer (*called Ba Moui Ba*) were often encountered along the road when our convoy would periodically come to one of these stopping points. An occasional watch salesman would also be seen about, regularly going from truck to truck, where he'd pull up his sleeve to reveal 7 or 8 wristwatches to those who might give him a little attention. They were mostly Japanese makes offered for sale at bargain prices. The only thing, though, for those who would foolishly purchase one, was that sometime later, when opening the back plate of the watch to inspect its inner mechanism, it was always discovered that its jewels had been removed from the working parts to further reveal it as worthless junk.

Our ten-ton 'Lowboy' trucks are shown here in transit with the plows aboard, stopped briefly near Di An to let the stragglers in the convoy catch up.

Winding our way in convoy, we passed through other villages with small rudimentary homes that often lined the road. It was interesting to note that some of them had flattened-out beer and coke cans displayed in an unusual, but resourcefully salvaged patchwork of aluminum siding. All were American-type beer cans, along with a few cans of Coke and Pepsi, which had likely been recovered from one of our military installations somewhere thereabouts. At that time, the process of aluminum canning was apparently absent from Vietnam, as all of their domestic beer and soda always came in glass bottles instead. But, it had struck me that, as the flattened aluminum cans had been tacked on and lapped over onto the sides of these little houses with the labels all showing, the combined mix made for a rather colorful patchwork, representing

Coors, Budweiser, Pabst Blue Ribbon, Olympia, and a few others. In a strange artistic way, these little houses seemed to take on a pseudo-Americana sort of look, while showing off their weird and wonderful inadvertent form of 'pop art'.

In the more rural areas, I would oftentimes see several sizeable dogs partitioned within some of the wire-fenced yards along the way, looking thin and underfed. It appeared to me that they were treated more like livestock than pets. I had heard stories about dog owners in these poorer areas who would sell their animals to those in search of meat. Fresh meat for the country's civilian population was a somewhat scarce commodity in Vietnam, especially in the poorer rural areas. The countryside was largely devoid of cattle, sheep, and other animals that we (*in the States*) are so accustomed to having as a meat source; although, some of the more fortunate people there would likely have a few chickens, and occasionally a pig or goat to raise, along with a fair prospect toward acquiring fish at an open market if they could afford it.

As far as I knew then, the Water Buffalo was the only bovine type animal in the country that possibly offered any similarity to beef, but being akin to oxen they were more valued for their strength and stamina. For that reason, they were nearly considered sacred by the rice farmers who worked them fairly often in the fields, and regularly used them to take their excess rice to market. They were also used to haul feed and other necessities in a simple, single-axle wooden oxcart. Noticing this most unusual profile of a large, horned, black water buffalo slowly pulling an old-world type oxcart along the open road tended to draw one's immediate attention. It was quite a striking contrast alongside more modern-day vehicles, and seemed so entirely out of place to me for the year of nineteen hundred and sixty-eight. But, that in itself, was just another remarkable aspect of the unusual landscape there, being simple as it was, within the third-world venue of South Vietnam.

In our daily interactions with each other, we had always used the unusual term, "Gook," to loosely refer to the Vietnamese people, no matter where we were. With its daily usage, we tended to be somewhat 'language lazy' and a little abusive in our general regard for the populace there. Most of our 'in country' forces had openly adopted this nickname and had used it quite freely much of the time, from as far back as anyone could possibly remember. As it was, I guess the name 'Vietnamese' was just too long of a handle, with too many syllables rolling off of the tongue. But, while many of us weren't quite sure of its exact meaning, we all knew that it wasn't a kindly term. Someone early on had suggested that it meant 'Foreigner', but, as it turns out, it was instead actually a lot more like calling a black man a 'Nigger'. However, as slang word usage was bandied about, it became a more common term in referring to them. Even though it was, by definition, a derogatory one, it wasn't always intentionally used in a degrading or demeaning way. Here's what the current American Heritage Dictionary gives, as a definition:

Gook n. <u>Offensive Slang</u> Used as a disparaging term for a person of East Asian birth or decent.

While some other interesting points about the country were later learned, it was also revealed that its national name, *Vietnam*, was originally a descriptive term, coined by the Chinese, with the first part, Viet, meaning 'foreigner', while the second part, Nam, meant 'south'. Both terms were applied together, to refer to those particular foreigners from the areas south of China. *{Perhaps, this is where the mistaken application of 'foreigner' got confused in the meaning of the word 'Gook'.}*

Once in a while, we would encounter a wandering tribe of Montagnards along the way (*pronounced, Montan-yard*), who were an indigenous, nomadic people, some of whom were still quite primitive in many respects. Their name, French in origin from the days of the earlier French military occupation, quite literally translates to *Mountaineer*. I recall that our convoy was stopped on one particular day for several minutes while a large contingent of them filed along and crossed the road in advance of us. In their basic mode of travel along dusty footpaths, this particular tribe appeared to be in serious need of bathing. Their primitive clothing for the lot of them was just a simple loincloth or shorts (*and no shoes*), with most of their children in tow remaining comfortably clad in their 'birthday suits'. More noticeably with this wandering tribe was the fact that its men and women wore nothing to cover their upper torsos. As they continued to pass by in review, many of them carried rudimentary backpacks filled with their acquired goods. They all took on a somewhat ghostly, pale white appearance from which the accumulated chalky dust along the side of the road had lightly covered their otherwise dark features from head to toe. Parading before us in full view, it was really quite remarkable to note that their primitive appearance almost seemed like a photo page right out of National Geographic.

These poor, native, tribal people were usually found in the higher elevations of the country and generally remained well removed from the Vietnamese populace. Although not all Montagnard tribes remained poor and primitive, many were still regarded on separate terms with the Vietnamese. In earlier times, small, armed conflicts developed between the two extreme cultural factions, somewhat like the skirmishes between our own Indian tribes and the American settlers. As a result of the Montagnard's mounting losses and dwindling numbers, the Vietnamese populated and worked the lower farmlands of the country, while the natives were rudely pushed out and reduced to taking their chances up in the higher, more mountainous areas of each region to roam about, as they foraged and hunted for food in their new element. With the earlier French and later U.S. involvements in the war, the Montagnard tribes had no part in the struggle between the two Vietnams and were generally left alone by most, but sometimes suffered occasional losses from incidental contact with North Vietnamese soldiers, or from mines and booby traps set by the Viet Cong.

On certain occasions in the field, a few Vietnamese prostitutes would also appear now and then, distracting us, as they aggressively solicited for their cause. They spoke just enough broken English to get their point across, even as some among them sported Betelnut stained teeth. Being virile young males of our species, there were usually a few amongst us who then felt physically enticed, and soon acquiesced to their driving hormonal animal urges to pay each girl five dollars (mpc) for a few minutes in the brush. (*This was commonly referred to as, 'Short Time'; not to be confused with the term, 'Short Timer', which was used to refer to those of us who were getting close to rotating out of the country.*)

From time to time, a few of the men would then catch Venereal Disease or Gonorrhea as a direct result of these unauthorized liaisons. In these instances, our medics would usually give out basic advice along with the usual prescription doses of anti-biotics. While this wasn't a sanctioned activity, our brass and higher-ranking NCOs were mostly unaware or tended to look the other way so long as it didn't interfere at all with the on-going operation. It seemed that, as long as this type of activity stayed out there and our brass didn't happen to catch anyone in the act, it was somewhat tolerable to some. But if caught and brought into the NDP or our Company area back at Battalion HQ, it likely would have led to more serious disciplinary action. Every so often, we were reminded by our Company Commander and platoon leaders that sexual diseases were more on the rise and running rampant among the prostitutes there. It was even suggested that the enemy may have been intentionally propagating the disease in order to create an epidemic among our own forces. (*I'm sure that assertion was more likely used as a scare tactic, as it always seemed somewhat absurd and farfetched to me*) But, in exercising their general concern for our wellbeing, they gave out the usual sound advice, telling us to simply exercise caution and try not to give in to our primal animal urges.

In further regard to this, I remember one particular unfortunate occasion, that involved a rather shapely young prostitute, who was carefully smuggled into our NDP one night. With the cooperation of a few members of our security, she was escorted back out, undetected, the following day. However, one of the men partaking with her that night, who was in the next tent over from me, had oddly developed an enlarged testicle just a day or two later, which suddenly posed a serious problem to him. While believing that it might clear up, instead it quickly grew in size to nearly that of a baseball. As his condition worsened, he couldn't help but exhibit signs of serious distress, while moaning and groaning with the pain from it. During the night, he just couldn't sleep, and neither could anyone else until it was finally brought to the attention of our medic, who soon had him on an outbound chopper for treatment. Throughout my remaining time in country, I never saw that man again, and only heard through our 'grapevine' that he was later removed to a medical facility somewhere in Japan.

Oddly enough, while attending one of our more recent Land Clearing Association reunions at Fort Hood, TX in the year 2000, I once again encountered our otherwise long-lost 'casualty of sin'. Although he had long ago fully recovered from his malady altogether, I refrained from mentioning anything about it, as there was really no point in revisiting what had obviously become quite a painful and embarrassing incident for him; one that had quite unexpectedly put an end to his tour as a land clearing engineer.

Riding 'shotgun' in a Lowboy truck with my plow up on the trailer behind, I commonly kept casual company with the driver, as our convoy moved steadily down the road. At that time, I could usually get the latest jokes and the general news about other happenings within the battalion and elsewhere, while we maintained our pace with the other vehicles. Of course, there were always a few drivers who just weren't very engaging or talkative. But, many others were. Occasionally, we would abruptly come to a stopping point, for whatever reason, when in and around small towns and villages. On one particular brief stop, we found ourselves parked right next to a Vietnamese take-out restaurant. The driver and I couldn't believe our good fortune as we immediately picked up the smell of spicy food cooking, with its savory enticing aromas wafting out in our direction. Chow time back at Bearcat was still about an hour away; as we were due to arrive sometime around 1700 hours. But, since we knew we'd be stalled there for a little while, we decided to take up our weapons and go check it out, rather than just sit there in the truck and salivate.

Upon entering the tiny restaurant, we looked up above the counter and noticed two separate menus on an overhead board; one for Vietnamese, and one for GIs. Of course, both menus were largely the same, but the prices on the GI board were significantly higher. This was just how some things were done there, as the people knew that we generally had more disposable cash than most of their fellow countrymen. However, as we were hungry, we shrugged that off and ordered up 'take out' items for each of us. I must say, it was indeed good. Along with the rice dish that I got were a bunch of cooked shrimp in a light, spicy sauce, which were very tasty and made for quite an enjoyable impromptu meal.

When our homeward bound convoy finally got into Bearcat late that afternoon, I offloaded my plow and stowed my gear back in the barracks while most of the others went and got in the chow line at the mess hall for the evening meal. Suddenly, not feeling so well, I stretched out on my bunk and thought to rest there awhile in hopes of recovering before showering and changing fatigues. When the other guys came back from chow, I was still there, but shivering and sweating with my teeth chattering uncontrollably. I laid there helplessly under the covers feeling nauseous, weak and disoriented. At some point, it got the guys' attention that I was seriously sick while they were casually playing cards on a nearby bunk. So they immediately gathered me up and carried me directly over to the battalion Aid Station.

At first, it was suggested that I might have had Malaria until, after the medic questioned me more about what I may have eaten earlier. After revealing that I had consumed some shrimp from a roadside restaurant, he immediately filled a syringe and gave me a shot in the hindquarters, which started relieving some of my symptoms in fairly short order. As I sat there, I was both relieved and dismayed, as it was indeed some tainted shrimp that had given me a good dose of food poisoning that day. From then on out, after experiencing that, I swore off all Vietnamese food. (*Although, I quite enjoy it today*) *As it turned out, the lowboy driver had gotten off easy in comparison from this little roadside indiscretion, as he didn't order up any of the affected shrimp, like I had.*

In early August of 1968, LCT would get a nasty taste of reality and come to realize the ultimate importance of maintaining close contact with our mechanized security forces. Two of our guys traveling in convoy with a wrecked plow aboard their low bed trailer had suffered a flat front tire on their 10-ton truck, near the village of Kien Tuong as they were hauling the plow back for major repairs. The mechanized security this time errantly misjudged the situation. Their track vehicles then proceeded down the road with the rest of the convoy, believing that the two of them would quickly change the tire and be along soon enough. Although the security kept them in plain view, it was from quite a distance that they realized their critical mistake. The driver and his 'shotgun' partner were completely taken by surprise by the VC's sudden strike, and wound up getting hit with small arms fire (*AK-47*). While changing the tire, they were left momentarily defenseless there by the side of the road, and neither one had a chance to escape the sudden assault. Subsequently, a satchel charge then detonated within the truck's cab, which rendered it to a completely demolished state. In the aftermath, the previously wrecked plow that they were hauling still sat solidly in place aboard the low bed trailer, completing the full grim ensemble of destroyed equipment and men.

The two killed in action were (*driver*) SP/4 John Schmude and Cpl. Ron Taylor.

* <u>John R. Schmude</u>   Age 19, from Pontiac, MI

* <u>Ronald B. Taylor</u>   Age 19, from Addison, IL

Date of casualties: 4 August 1968

Panel 49 W ~ Line 10 and 11 *

* *Both Schmude and Taylor are listed on the 'wall', right next to each other.*

Sadly, I remember that Schmude had owed me fifty bucks from several months back, and that he and I had repeatedly argued about it over the course of time. In his frustration with me for pestering him about it, he refused to pay me back and I had no recourse but to take him in front of the CO, who then ordered him to take care of it by the next payday. He never did get around to paying it off, though. But, of course, after that ill-fated day, it really no longer mattered.

While it was always in the back of my mind and in the minds of others there as well, I'm sure that death or serious injury was a reasonable possibility for any of us at any time, given our wartime occupation and the notorious areas we labored to clear. We all fully realized the dangerous nature of our daily tasks, as we were effectively stealing the VC's cover, and they clearly didn't like it. On occasion, their response to our tactics would wind up erasing someone from our ranks, leaving us somewhat numb and with a lingering hollow feeling for a time. In total, six men in my company paid the ultimate price for their Vietnam service during my elongated 18-month tour there. In separate incidents, the VC had victimized five of them at random with RPG and small arms fire, while one died from accidental causes; all in the course of performing their appointed duties. Although it happened to be those particular six, we all knew it could have easily been any of us.

A final tally after the war brought the total amount of men killed in action to 7 within the full active term and existence of the 86th LCT & 501st LCC (*as both designations were the same unit*) from its original LCP formation in mid-1967 to sometime in April of 1970, when the equipment was given over to an ARVN Engineer unit and the company was dissolved; its men being disbursed to other units in country or sent back Stateside.

Those seven men were:

Sp4 Clarence E. Dalton Jr. Age 19, from Rollinsville, CO	86th
Sp4 John Robert Schmude Age 19, from Pontiac, MI	86th
Cpl Ronald B. Taylor Age 19, from Addison, IL	86th
Sp4 David O. Hershiser Age 20, from Colorado Springs, CO	86th/501st
Sp4 Jerry R. Davis Age 21, from Columbus, MS	501st
SGT E-5 Jimmie Jack Jones Jr. Age 20, from Steelton, PA	501st
PFC Paul V. Quaglieri Age 21, from Burbank, CA	501st

Cpl. Ron Taylor, slogging across the NDP motorpool area, while headed for his 'shotgun' seat in the Lowboy truck.

In addition to the men who were killed in action, there were also those who periodically became wounded to some degree. Most of these men suffered minor wounds and returned to action shortly after their injuries healed; a few of them had even sustained multiple flesh wounds on occasion. However, an unlucky few suffered even more serious physical consequences from their particular enemy encounters, or from the occasional accidental mishaps that sometimes occurred. These men were quickly evacuated out by chopper to field medical hospitals whether in country or in Japan, prior to being sent back to the States where they would later receive a medical discharge, given the overall status of their debilitating injury or disability. Those few unlucky men we simply never saw again, and their sudden loss from our ranks was genuinely heart-felt as well.

Operating in the field for a time without a host battalion, we had come to see ourselves more as a sort of maverick-type unit, to some degree. We were actually getting accustomed to surviving while procuring our necessary field supplies through our ever-helpful security elements, until sometime in November when our unit came together with the 168th Engineer battalion, out of Di An, for a brief period. It was at that time when we joined together with our former rivals, the 27th LCT, under the banner of the 168th, for just a short time frame of about two months or so. After that, in January of 1969, both the former 86th and 27th Land clearing Teams changed designations again, while severing attachments with the 168th to join the newly designated 62nd Land Clearing Engineer battalion, in Long Binh, and subsequently became the 501st Engineer Company, LC, nicknamed the 'Rome Plows'; with the 27th likewise changing their numerical distinction to the 60th Engineer Company, LC, as they kept their nickname of 'Jungle Eaters'.

(*Our company nickname prominently appeared on our dress greens in a red and white arched form of lettering, just over the castle on the red & white 20th Engineer Brigade patch. This unit patch was also worn on our jungle fatigues, but in subdued form, along with our rank.*)

This unique new Land Clearing contingent, in its initial formation, also created a new LC company to give the innovative battalion 90 tractors in all, by which to operate with. It was designated as the 984th Engineer

Company, LC, and they proudly sported the nickname of 'Land Barons'. This brought the regional forces of Land Clearing together for a much needed, more organized and coordinated effort. It made it possible to further serve the security interests of all, mainly within the III Corp Tactical Zone, and provided better logistics and maintenance support from within. A few months prior to this new consolidated endeavor, LCT had vacated the old home base of Bearcat and moved northeast, (on QL-15) about 10 miles up the road, to Long Binh Post, where we'd later see it become our new permanent home base.

Over time, we occasionally received some sporadic in-country news reports from our Officers and higher ranking NCOs. From these reports, we learned that there were other Land Clearing elements out there in the other tactical zones, which operated much like us, and possibly even suffered more casualties than we did in the course of their own operational activities. It was pointed out that the Marines in I Corp had their own Rome Plows; and in other areas, there were also a few smaller groupings of land clearers, with only about 5 or 6 plows regularly working the cut. Otherwise, there were two or three other full, Army Land Clearing Teams established in different areas within the zones of I and II Corp to the north. Some of the other Army Infantry units oftentimes employed just a single plow and a bull blade to clear around fire support bases and carve out spots for LZs, to satisfy their own basic needs and security concerns within their particular areas of influence. We had also learned that the Rome Plow wasn't the only jungle clearing piece of equipment out there. There were also a few other units who had a combination of Rome Plow tractors and one or two 'Letourneau' Tree Crushers as well. These 60-ton tree crushers were monster machines that could fell large trees, while splintering and pressing them into the ground as they went. Unfortunately, as impressive as they were, they're presence didn't last long in Vietnam. The enormous size of this machine had always made them an easy target, and its overall weight factor often made them quite difficult to transport from place to place without dismantling them.

However, the latest news happened to reside with us here in III Corp, where our own group of Land Clearers had come together to become part of an even more unique assemblage. While forming our newly consolidated battalion in Long Binh, it was found to be the first time that any of the land clearing teams was physically united in their common cause, serving as one collective and cooperative unit. Prior to 1969, there simply weren't any combat engineer battalions which were entirely comprised of land clearing teams and their direct support elements. From our company's standpoint, at long last, we were back within a battalion that could provide all of the necessary support that was called for, as we continued with our never-ending line-up of land clearing missions. As a result, we no longer needed to rely on our mechanized security elements for help on requisitioning parts and supplies. For a change, we then tended to get the items that we really wanted instead of simply having to settle for whatever might have been available to us.

Being from one of only two existing Land Clearing Teams within the entire III Corp zone, we once again found ourselves paired up within this new battalion with the old 27th LCT (*in the form of the 60th*). Earlier, while in the 86th, we had always heard different reports about them regarding the various amounts of cut acreage that they were achieving, as well as their operational incursions with the enemy, and their somewhat grim casualty reports that generally followed. For reasons unknown, their casualty rate appeared to be much greater than ours. They seemed to get hit by the enemy a little more often than the incidental incursions that we had been experiencing. Because our brass only had the 27th by which to compare us with early on, they had somehow become somewhat of a rival unit to us by which we could competitively measure ourselves and try to juice-up our efforts, toward clearing out more acreage in a shorter time frame.

This is a partial view of the 501st's motorpool on Long Binh Post, taken from atop another plow. It shows a few D7E dozers and their parts in disarray, as they were set aside either for re-building or for cannibalization. More recognizable out in the background, a couple of our bull-blade dozers were parked with one of our air compressors openly sitting out in front of them. Below is another angle of the motorpool in Long Binh, with our Company area partially shown, sprawled out in the background to the extreme left. One of our M-548 Maintenance Tracks can be seen at the center of the photo.

Settling into our new battalion area on Long Binh Post, we found it to be much larger compared with our former accommodations in Bearcat. Our new Land Clearing battalion had secured us a larger company area and a bigger outdoor movie theater, as well as a sizeable PX (*Post Exchange*), EM Club, NCO Club, motorpool...everything just seemed to be on a larger scale there. Our new living quarters were in the form of single-story type billets with metal siding and concrete slab floors, which featured a 4-foot exterior wall of sandbags that surrounded the buildings for marginal bunker-type protection in the event that a rocket or mortar barrage suddenly impacted nearby while we slept.

Our Engineer battalion's newly re-designated numerical distinction of the 62nd seemed a little strange at first, after being quite well known as the 86th for so long. Although our official status had been changed from a 'team' to a 'company', we were still widely known as LCT. Only then, we had become the 501st LCT.

~ Twelve ~

Exiled

Sometime between LCT's separation from the 86th Engineers and the new Battalion reorganization in Long Binh, I got in a little trouble one day out in the field that netted me an immediate transfer out of the unit. While cutting in quite hilly terrain around the Bien Hoa area, my plow suddenly went up on an unseen stump with its right track, just as it simultaneously sank into a depression of sorts on the left side. Immediately recognizing the sudden tilt to the left, I quickly locked both brake pedals down on the floorboard, while disengaging the transmission, which abruptly halted the plow's forward progress. But, my response was a bit too late. The unexpected tipping motion raised the right side of the heavy tractor just enough to ever-so-gently tilt it up on its side, to a point of apex, where it amazingly remained suspended in air for what seemed like the longest time. Incredibly, the plow just stopped there, in mid-motion, poised unwavering in its newfound position. As its right track angled up skyward, it gave me a very long and fretful pause in which to take it all in. While I anxiously sat there in full tilt, it was almost as if, suddenly, time itself had frozen for the moment. The tipping motion's upward momentum had ever-so slowly arrived at a standstill after easing the tilted tractor to a near perfect balancing point, where it unexpectedly remained as if it was simply leaving it all up to me to decide its downward fate.

Cursing in disbelief, without a recourse measure for this sudden, unforeseen situation, I could only sit there and face the overwhelming feeling of sheer helplessness, while awkwardly positioned sideways within the operator's seat. At that point, this weird and unexpected surprise had succeeded in getting my full, undivided attention, as I endured the anxiety of the moment from this extreme near-perpendicular lean. I just didn't know and couldn't really determine exactly which way the weight of this teetering mass of steel might finally shift; toward either rolling the tractor over onto its left side, or toward possibly finding its way back down onto its tracks to properly right itself. However, despite this frustrating dilemma, I was still rendered powerless to act, while simply waiting for the tractor to keep its imminent appointment with the earthly laws of gravity.

Anxiously getting up from my seat, I was unwilling to resign myself to this uncertain fate, and foolishly tried to counter-balance the plow in a desperate attempt to get it to right itself. In trying this, I frantically leaned against the hydraulic tank, just to the right of the seat, only to soon realize that my meager weight wasn't about to influence it in any way. Then, with a slight degree of movement detected, I could see that it had suddenly shifted out of balance, and was indeed going over. Clumsily falling back across and down into the operator's seat, I hung on tight, bracing myself for the inevitable. As the plow slowly rolled over, I managed to hang onto the armrests of the seat, but wound up banging my head against the thick steel-meshed screen on my left. The heavy tractor impacted somewhat violently and slid down the hill, where it eventually settled on its left side, lying partly on its top, as it finally came to rest within the base of a small gully. Aside from sustaining a lacerated ear and being somewhat disoriented from the blow; while assessing things from my nearly upside-down position, I believed I was ok. I then hastily scrambled out without giving any thought to my rifle, which remained where I had always secured it, stowed vertically behind the operator's seat.

Standing atop a small hill, with my overturned plow in view below, I was still feeling somewhat dazed and disoriented from the violent blow to my head. Then I suddenly heard a strange whirring sound overhead and turned to look up and see a B-40 rocket heading in my direction. Oddly, it seemed rather slow for this type of shoulder launched rocket. Its flight path was also noticeably too high and I could tell right off that it would overshoot the plow and myself from where I stood. But instinctively, I ducked down anyway, anticipating its imminent blast. It then buzzed overhead and consequently entered the sloping wooded jungle just beyond where my overturned tractor lay helplessly on its side. However, aside from the earth-softened thud that came from the rocket's impact, there was nothing else to be heard. Quite surprisingly, and much to my relief, despite its lethal appearance, it turned out to be a big dud.

I had initially suspected that there must have been some particular malfunction with this rocket, because of its loud buzz and unusually slow rate of travel.

Somehow, I managed to locate my camera, to get this downhill shot.

Fully realizing what had just occurred, I quickly stood up and looked up range, across to the opposite woodline area from where it came, and spotted a half naked, older Vietnamese man, dressed in some sort of lower waist wrap, like white swaddling cloth. He was just standing there looking back at me, appearing quite skinny, as his boney ribcage stood out in clear, definitive form. Oddly, it almost appeared as if he was wearing a big white diaper with the way in which it was wrapped around his waist and loins. As he intently stared back in my direction, I too kept a steady fix on him. Lo and behold, it was none other than *'Charlie'* himself!

Our frozen mutual gaze seemed to last for the longest time, until he finally turned away and darted back into the jungle, quickly vanishing within the thick, leafy, woodland type cover that so easily facilitated his hasty retreat. That turned out to be the single most haunting moment of my Vietnam tour, as it was an actual face-to-face encounter, although at a distance with one bent on killing me. It often made me wonder how he may have reacted from that distant vantage point after missing the mark and then seeing that his projectile was a dud. I also can't help but wonder what might have occurred if my plow hadn't taken its fortuitous spill, which seemed to have changed his initial plan a bit. He was surely waiting for me to come into closer range before that weird tip-over event occurred. I suppose his extended view of the bottom side of my tractor, noticing it helplessly angled up skyward, may have astonished him as well. He had been patiently waiting for me to give him a nearer window of opportunity before getting off a more accurate shot. If I *had* been in closer range, would it have detonated, had it made solid contact with my plow? And, with the rocket's errant departure rendering him virtually weaponless except for the launcher, did he notice that I, too, was without my rifle, as it had erroneously remained behind the seat, inside my overturned plow? These nagging, unforgettable images and questions oddly lingered on in me, living within the depths of my memory throughout the many years that followed. The weird, spooky 'tip-over' incident would sometimes revisit and even preoccupy me, during some of my quieter moments of personal reflection. In a surreal moment of sheer happenstance, it all seemed to reveal a very strange and unusual twist of fate that still continues to haunt me to no end.

Within minutes, two other plows arrived on the scene; one with our new young 2nd lieutenant hitching a ride aboard. As he came into view, he looked to be a new addition to our unit, as I had never encountered his presence prior to that day. While acknowledging him, I couldn't help but note how young and boyish he looked, as our company seemed likely to have been his first assignment following graduation from high school and OCS. Noticing my overturned plow lying helplessly down the hill, he immediately inquired as to what all had just happened there. After relating the details to him, he ordered me and the other two plow operators to grab our weapons and follow him on foot into the woodline, to the very spot where the failed enemy combatant (*Charlie*) had disappeared, and go conduct an immediate search for him. (*Search for him?*) Taken aback with that, I suggested that we should instead contact our mechanized security, who were somewhere just to the rear of us, and let them take care of it. Agitated, he remained adamant with his wild proposal to go in after this guy. Quite plainly, I thought he was crazy, and told him so. Throughout my time in the Army, I had always had a healthy respect for my superiors. But, this went too far outside the realm of common sense and, although he gave me a direct order to go along on this, I refused. After all, we were Engineers, not Infantry, and not fully trained for that type of pursuit activity. Plus, none of our other officers and NCOs had ever suggested anything like this before, to my knowledge. Granted, we were soldiers first, although not well schooled in the ways of the enemy like our mechanized security was who just happened to be only a short distance behind us. This all made perfect sense to me, but, in our interactions with our officers, we weren't supposed to give any particular thought to their orders except to comply. We had been instructed from 'day-one,' to simply follow them, without question.

That would be all that I needed, after a close call like that: some raw 2nd lieutenant leading us into something totally uncertain and completely unnecessary, where we might have drawn unseen fire and suffered whatever consequence that might have otherwise befallen our tiny detail of engineers, while tromping through the thick brush in search of an elusive and cunning Viet Cong guerrilla.

Going into the jungle in pursuit of this lone VC just seemed like a real dumb thing to do, especially when we had mechanized security that were assigned to us for that very purpose, even though they happened to be a bit late in responding to the particular incident. But, aside from that, our mechanized security clearly wouldn't have handled it that way, with only three or four men on the hunt. If anything, that 2nd 'Louie' should have thought more about his own men and backed the plows up until the security caught up with us. That's the procedure that I had always known and practiced from these types of incidents in the past. But, the two other operators resigned themselves to follow him on his mindless quest, only to turn up nothing and, fortunately, they all returned without further incident.

This young 2nd lieutenant, in his frustration, had promised to report my apparent insubordination, and I then found myself later trying to explain things to the CO. But, he would hear none of it and found that the best overall solution was to just get rid of me. Consequently, I soon afterward found myself headed down into IV Corp area, to the somewhat small locality of My Tho with new orders transferring me to B Company of the 86th Engineers. There was no 'Article 15' with a one-month partial reduction in pay, nor was I busted down in rank or confined to the company area, or any other such common military type punishment. He simply wanted to wash his hands of it and banish me from the 'team' for not following orders, no matter how ridiculous they may have seemed. Although I had expected to be punished for my apparent transgression, I never really thought it might result in full banishment. To me, that was actually more humiliating than losing a stripe or two. But, there was nothing that I could say to effectively reverse my unexpected plight. So, I then dejectedly left the CO's tent, suddenly feeling a degree of shock setting in.

Somehow there was little satisfaction in later knowing that this young 2nd lieutenant also got the 'boot', and was reassigned to an Infantry outfit within the 9th Division. I imagine, in his case, there was likely a serious lack of confidence found in his overall ability to reasonably direct his subordinates while trying to maintain a heightened perspective on safe practices, as the men handled the equipment in the field.

In reluctantly departing our NDP, believing that I had seen the last of my days with Land Clearing, I returned to Long Binh, where LCT had been set up in temporary housing for their periodic 'stand downs' after becoming estranged from the 86th. Upon receiving my transfer orders, I headed directly southwest, along QL-1 highway into the IV Corp tactical zone, riding in the back of a Deuzenhalf truck. The truck then progressively wound its way down through numerous towns and villages, toward the little riverside hamlet of My Tho. After traveling about three hours out of Long Binh with my gear in tow, I finally arrived at the newly established home of the main body of the 86th Engineers. It was located within the northern portion of the Mekong River delta region, where their battalion compound was neatly situated along the bank of the soft flowing My Tho River. The northeast side of their compound there was flanked by a small airstrip that served the interests of the battalion. The southwestern edges of the base skirted the bank of the river with a one-lane vehicle bridge leading out. This small bridge spanned out across the greenish, lightly rolling watercourse to serve as a main access point for the daily movements of our own security forces on the opposite side of the river. Additionally, it served to facilitate the daily flow of local post workers from their homes, just out beyond the nearby woodline.

My newly assigned job with B Company was to essentially tinker around in the motor pool there, simply helping with the maintenance on their two bulldozers since they already had skilled operators assigned to them. These versions of the D7-E were the standard, construction type with regular bull blades attached and no protective cab. Because I had spent so much time working in and around Rome Plows, and kept their familiarization fresh in my mind, I came to see their two standard dozers as suddenly looking naked and incomplete, seemingly out of place there. But, in reality, *I* was the one who was actually out of place, somewhat like a fish out of water.

Believe it or not, it was actually somewhat difficult for me to get used to that slow-paced level of activity there in My Tho, with not a lot to do in their motorpool and nothing halfway stimulating to take my mind off of my remaining time in country after spending so much intensely active time within LCT. My job just entailed basic maintenance and part replacement for the two dozers and some of the trucks. The leisurely-paced humdrum feel of stateside duty set the tempo for each day, while B Company's Construction crews worked on a daily basis, in repairing and resurfacing the various nearby roads and airstrips. For much of my time in country, I had been more accustomed to having an important objective to focus on, while simple equipment maintenance, although important, just didn't quite do it for me. However, from B Company's position, that was simply all that was available to me, at that time. Of course, if they had given me one of those dozers, along with a series of construction related jobs, I might have more easily acquiesced to my exile from Land Clearing, and switched off the funk I was in. But, at best, I had become a reserve operator there, without an assigned piece of equipment to raise my enthusiasm, or to even help further warrant my AIT training. Clearly, my overall experience with Land Clearing operations was of little or no value to a construction company, and I think they may have even been somewhat at a loss, as far as knowing exactly what to do with me.

Obviously, no one else there shared my lack of enthusiasm, as they all seemed somewhat content with the cards they were all dealt. But, I made a few new friends there in My Tho, after settling in, and had more time to regularly go to the PX and EM club, as this banishment seemed more and more like stateside duty. As time went on, I had occasionally voiced my bottled-up discontent, in the company of my new-found friends, while living within the squad-sized hooch that we shared. They tried their best to reassure me, in suggesting that I had a much better shot at surviving the war, if I just settled in and learned to like it there, despite not having my own piece of equipment to run. Furthermore, they reiterated that it wasn't nearly as difficult or hazardous as the grueling job of Land Clearing had been. These were a great bunch of fun-loving guys, who liked to joke and kid around a lot, and who took a particular liking to me, in spite of my ongoing discontent with what I was left with there. Although they couldn't fully understand why Land Clearing had made such an overwhelming impression on me, and why I couldn't just accept my good fortune of soft duty, collectively assuming that my basic reasoning abilities had somehow gone askew.

Relaxing in my little section of the hooch, in My Tho.

Given that there were no restrictions on alcoholic beverages in My Tho, unlike when I had worked in the field with LCT, many of the guys in my hooch tended to stockpile cases of beer and hard liquor, while establishing makeshift open bars within their small, sectioned off living quarters. This was much like how things were back at Bearcat and Long Binh, when everybody came in on Stand Down; although, in this case there was no end to it, with many of the guys there, getting drunk most every night, except when called upon to pull perimeter guard duty. I occasionally found myself pulling guard duty there, while stationed within an elevated sandbagged bunker, out along the river's edge, for two hours of night watch, with another one of our company's men. Inside the bunker, we had a direct COM line with our security forces there, by which we would report any particular movement that might be occurring, out within the open area across the river, while occasionally calling for illumination flares, as we continuously scanned the nearby landscape for any sign of enemy activity.

Otherwise, organized volleyball games were regularly seen after hours in the company area, with movies and half-drunken card games often occupying the evening hours for many of us, unless of course, a somewhat rare mortar attack occurred to break things up for a while. I even had time to go to the dentist while there, traveling several miles northwest, to Dong Tam, where the 9th Infantry boys were set up.

Christmas time, with the guys in my hootch. (My Tho- 1968)

It was really a much more leisurely assignment, being relegated to serve in one of the line companies of the 86th... but, under the conflicting circumstances, I came to hate it. Some others would have preferred it to the hazardous, gritty-grimy job of Land Clearing, and would have likely been overjoyed with the disciplinary transfer that I had received. However, at least with Land Clearing and all of its arduous activity, time seemed to fly by, while with B Company, time simply dragged. In my own frustrated rationale, I had come to feel that time served more actively there, seemed to be the best mode in getting me through this tour, while ultimately keeping me sane. But, besides that, I had always enjoyed the outward adventure that regularly went along with Land Clearing, although occasionally somewhat hazardous, while clearing jungle in various terrains, and while moving about from place to place, to see a bit more of the outlying countryside.

Additionally, I had always felt that my efforts with a Rome Plow were largely worthwhile and even vitally important, as this key aspect of saving lives through the use of heavy equipment, had somehow appealed to me, and had drawn me in, in an odd sort of way, despite the nagging elements of danger and hardship that normally went hand-in-hand with much of the work. Plus, in sticking with my near-obsessive time management theme, it had always kept me thoroughly occupied, active, and somewhat well focused.

Our holiday spirits were proudly displayed. (My Tho-12/68)

As December and the holiday season rolled around, many of us turned our thoughts toward home, and felt the loneliness of isolation from our families more then, than at any other time. With Christmas nearly upon us, it was announced, through newly developed technology, and the generosity of a commercial communications company named, ComSat, that it was possible for us to call home and actually speak with loved ones.

We were then living in a new age, regarding communications, as satellites were being used to relay telephone conversations around the world. Prior to this, it was possible to talk through short wave radio over long distances, relaying the signal over land and sea, to a short wave station that was able to re-transmit it from there, as ships in the Pacific had performed that task in the past, by patching conversations through, to span a call over limited distances. But, with the U.S. response to Sputnik, which was the very first satellite launched in space by the Russians, TeleStar had become *our* first communication satellite that bounced short wave messages from far off places, over even greater distances. After ComSat had later purchased TeleStar, and launched another satellite, it had become increasingly easier to bounce conversations back and forth without further transmissions along the way, as they relayed them from satellite to satellite, until the signal would ultimately reach its intended destination on the other side of the earth.

As it applied to us, it just so happened that a ComSat contingent was traveling around different areas of South Vietnam, just as the holidays approached, purposely offering this new communication system to servicemen and women stationed there…for free, while they tested and demonstrated its wireless, long-distance application. They had a series of four or five modern big-rig trucks, featuring futuristic circular space age logos that were emblazoned on the sides and rear of their boxed trailers, with a large satellite dish mounted atop another trailer, which was the main distinct feature that gave them away, as they pulled into our compound in My Tho, and set up shop. As I recall, the satellites had a certain 'window' of opportunity, and it was figured that we could each be allowed five minutes for our family conversations, while the orbiting COM portal was overhead, as we all lined up and took our turn in the booth with it, saying "over", at the end of each transmission. As I recall, they had about 3 or 4 private booths, and were able to use them all simultaneously, to accommodate everyone as expeditiously as possible. Our conversations were apparently still broadcast over short wave radio, even though we were using standard telephone handsets, but the vibrant sound of my parents' voices still came in loud and clear, much like with a local telephone call. The other men still working out in the motorpool area were also summoned to where ComSat had set up, as our CO wanted to make sure that everyone in the unit had ample opportunity to make their Christmas call.

Sometime just after the turn of the new year, I suddenly got word from home that my mother had suffered a heart attack. To what degree, I didn't know, as my father just gave me the basic information, and didn't want to worry me. Although she survived it, I didn't quite know how potentially life threatening it might be, so I went to see my CO, and asked if I could take an emergency leave. He said it wasn't the sort of thing that qualified as an emergency, in order to warrant a 30-day leave. But, as I emerged from talking with him, our chaplain approached me, as he had overheard our conversation, and suggested that there was possibly another way in which I might get that 30 day leave. He explained that if I were to voluntarily extend my tour for an additional 6 months, it would qualify me to take that leave right away. Plus, upon my return, I could go just about anywhere in country, as far as a new duty assignment, while keeping within the bounds of my MOS (*Crawler Tractor Operator*). With all of this new information now on the table, it no longer mattered that I was getting 'short' on my original 12-month tour, having only about 50 or so more days remaining, at that point, before being due to rotate out.

I could now actually write my own ticket, and get the hell out of My Tho!

The guys I lived with in B Company thought I was crazy to want to extend my tour and return to Land Clearing, as they knew of the perils associated with operating Rome Plows, along with the overall harsh conditions in the field. A few of them tried to further dissuade me from it, with their well intentioned words just falling on deaf ears. For me, it was a personal vindication of sorts, to have a chance to right the wrong of my unfair exile from Land Clearing, and pick up where I had left off. Essentially, I regarded it as unfinished business that needed to be rectified before rotating out, and moving on to my next duty station. Otherwise, it would gnaw at me forever.

So, I said goodbye to my friends in B Company of the 86th, and after turning much of my issued goods back over to the supply sergeant, I caught a ride up to Ton Son Nhut Airbase, in Saigon the following day, with nearly a full duffle bag in tow.

Arriving back home, in northern California, after the long flight from Saigon, I found my mother to be stable in her condition, as it was diagnosed as a light heart attack, and not life threatening. While relieved of that worry, I spent most of my leave bumming around with family and friends, essentially whiling away the time and having a little fun. Before I knew it, my 30 days were up again, and this time I found myself flying out of Travis Air Force Base on a plane loaded with fresh Marines, who were already fully outfitted with their camouflaged jungle fatigues, as they were primed and ready, while anxiously anticipating their new, unfamiliar tropical environment. I was still dressed in 'civvies', and stuck out like a sore thumb next to all of them, but I didn't much care.

The flight back was the only one I could get that could possibly put me back in country without running the risk of becoming AWOL, as 'Military Standby' was the sole form of air travel available to me then. The only thing was, this plane was headed for Da Nang, which was quite a far distance north of Saigon and Long Binh Post, where I was due to re-join my former unit. Upon arriving in Da Nang, I changed into my jungle fatigues and put in for the next flight to Saigon. After about an hour's wait, I managed to catch a C-130 transport plane, for the 2 to 3-hour trip south. These big, four engine turbo-prop driven cargo planes were not at all like the commercial jets, as the flights were slow and noisy in comparison, and the standard seating was troop style, with long bench-like nylon mesh seats running along the length of both sides of the fuselage. This particular type of aircraft was actually designed for hauling heavy cargo as well as troops, which in this case, was in the form of multiple palleted, boxed goods that filled a section of the cargo bay, just aft of the seating area. The view, whenever I got up to look out, often featured seemingly endless, thickly forested jungle below; an environment that I was already well acquainted with, and headed back into.

From Saigon, I finally found my way over to Long Binh Post, as the truck I was in, passed by the post's main amphitheater, where we noticed while driving by, that it was being readied for Bob Hope's USO Show, which was scheduled to take place on the very next day. I had often seen TV footage of some of his earlier U.S.O. shows, and thought; perhaps, it might be possible to actually catch this show, given that it was only a day away. But, as it was, I reported in, to the 501st Engineer Co., LC, and with their explicit instructions, I promptly took a chopper out the following day, to re-join my old outfit in the field, as I wound up missing the opportunity to see that show. But, it was exhilarating to be back out in the field again; and although things had changed, as far as the new battalion status and such, it was still the same old LCT, with the same look, only with some new faces to go along with the other familiar ones that I had earlier come to know. Additionally, with the new battalion and Company status, we then had a new Company Commander in charge of us. (*Captain Jack MacNeil*)

Some of the guys were fairly surprised and bewildered to see me back, as they had previously figured that I was likely gone for good. But, in restoring my sanity, and renewing my commitment to land clearing, I jumped on a plow the very next day and resumed as if nothing much had occurred. I was now back in my element, where I could re-apply myself to the task at hand, while convincing my psyche that the additional time remaining in country would move along somewhat quicker as a result, and before long, the end of my tour would be within sight. As far as I was concerned then, all was forgiven, as my short term in exile had been served, and my subsequent reinstatement was complete.

While I was away, a third plow platoon had been added to the company, effectively dividing the equipment up three ways, with each platoon then down to nine plows and a bull blade. Also, the numbers had changed on the plows, as we now had the series ranging from 60 through 89, and I was given Hershiser's old rebuilt # 83, which lasted for only about a month or so, until the engine finally lost significant power and blew. I then got # 87, and operated it for quite some time, as I resettled into my old outfit, within the newly added third platoon.

Upon my return to LCT, I had learned that David Hershiser, who earlier shared our squad tent, and operated plow # 83, had taken in an accurately fired RPG round, one day while I was still exiled down in My Tho, as it entered his cab and detonated.

He was 20 years old, from Colorado Springs, CO.

<u>David Owen Hershiser</u>

Date of casualty: 4 January 1969

Panel 35 W ~ Line 34

While on an operation northeast of Saigon, in the Song Be area, we were cutting in mostly dense woodland type jungle, with some areas of thick Savannah grass and a few patchy open spots on the landscape. In this area, the tracts that we cut out were much larger than the normal roadside cuts of 200 – 300 meters in depth, as they were considered to be part of an enemy stronghold area, that our security had particular interest in, toward getting a better handle on the out-of-control insurgency thereabouts. When I finally turned a corner and came upon a clearing, I noticed a strange dark-reddish, dome shaped structure, just off in the distance, with a long, high chimney rising from it, which took me just a moment to realize that it was a large brick kiln, that had been constructed for firing ceramics. While ever the curious one, I pulled away from the cut, and crawled my tractor over to this kiln site for a bit of a closer look. Noticing that it had sustained serious damage from apparent enemy activity, I could clearly see that the kiln was compromised and rendered entirely useless, while having long been abandoned.

However, just to one side of the ruins, was a rather big pile of glazed ceramic vases; all being of very large size (*approximately 36" high*), with wide lipped tops, and all were very similar in shape and composition, except for their particular variations in color. As I looked closer, I found that some were simply plain, from being fired in one solid color, while others had decorative designs and a combination of colors, with rope loops having been established at the base of each neck, suggesting that they were likely intended to serve as hanging water vessels.

Jumping down from my plow to approach the pile for a better view, I could see that most of them were partially broken or chipped, and not really worth bothering with. But, while looking about further, I picked out a shiny black one that didn't have a mark of damage on it, and set it aside. Then, I found another one, fully intact, that was much more colorful, as it featured a dragon circling the belly of the glazed vase, with the surrounding areas cast in tan and greenish-blue. I thought it was so nice, that I took pains to lift it onto my shoulder, and slowly get it up and into the plow's steel cab, while carefully setting it down between the hydraulic tank and the seat, as I then somewhat gingerly crawled the tractor back around, and away from the cut, to where our Deuzenhalf was parked. With a little help, I got it out of the cab and up into the back of the canvas covered truck, where several sandbags secured it for the short ride back into our NDP. Although, since it was actually perfectly secure right where it was, it just remained in the back of that 2 ½-ton truck, positioned safely out of harms way, for the duration of that field operation.

When we finally came off of that operation, and rolled back into Long Binh, I didn't really have a plan as to where I might keep it, as space was limited in the billets, and it was very fragile, as it was. But, after casually mentioning my acquisition of it to our supply sergeant, he offered to store it, and take good care of it for me, right there in our supply room, provided I didn't mind that it might be used as a receptacle for his cigar habit. So, as promised, he kept it there on display, right next to the counter, for the whole company to see, where it served as a giant ashtray, until my time dwindled down to about a month shy of rotating out.

I remember, at some point, as the days drifted by, that one of my 1st Sergeants, 'Top' Willard, who was from Tennessee, had somehow acquired a small dog while there in Long Binh, and let him have full access to the entire company area, as he would come around and visit the various hooches and alternately make his rounds over to the barracks to mark off that section of his territory. He was a cute little tan colored terrier mix of some sort; but since we never learned that Top had named him "Poor Folks", some of us just mischievously referred to him as "Bastard". I imagine Top may have rescued him from an uncertain fate there, while they soon became fast friends and a notable duo around the company area. Occasionally they were even seen together, presiding over one of our daily formations on post, as it was the First Sergeant's task to run the business of the company, whether he was out with us briefly on another field operation, or back at our company area in long Binh. The contrast with these two was remarkable, if not a little comical, with this little dog occasionally sitting beside our tall, handle-bar mustachioed First Sergeant, as he stood out in front of our formation to conduct the usual business of the day.

The little dog turned out to be a devoted sidekick to him while remaining in and around the battalion area; but when Top's time came, to rotate back out to the states, he had to give him up, as he couldn't take him back home with him without incurring a lengthy quarantine period. Exactly where that little dog went, though, nobody really knew for sure; perhaps to someone in one of the other companies, as none of us ever saw him again, after Top left.

~ Thirteen ~

Shepard's Way

While some of us were more adept at cutting down trees, during the course of our field operations, others attempted to acquire that skill by practicing on available timber whenever possible. From my earlier days in the 86th's version of LCT, most everyone had come to recognize that a guy named Shepard (*out of 1st Platoon*), was probably the best that any of us had seen, when it came to skillfully carving the trunk away, and getting the tree to always fall in the desired direction, out of harms way. He just had the knack, and maintained a keen awareness of where he was, and where he needed to be with the plow at all times; and it was always a pleasure to watch him work whenever the opportunity presented itself. There were a few other operators among us who were actually pretty good at it as well, while still others struggled to get the rigorous technique down pat. But, from all those whom I had seen, he was the standard in my mind, by which everyone else could be comparatively sized up and measured accordingly.

With a variety of sizeable trees available within these thickly forested jungles that we encountered, a few distinct types stood out and tended to offer more of a challenge to our acquired tree-cutting skills. Teak trees were somewhat common in many areas, with some very large, towering ones found on occasion. Their thick, dense trunks were a little tougher to split with the sharpened 'stinger', as this particularly dense hardwood tree usually gave pause for additional caution, while having to make more stabbing passes at it, until it cracked and began to lean to one side (*hopefully, the side where it was projected to fall*). When addressing hardwood with the stinger, the operator had to gauge just how far inward to aim it, from the outside edge of the tree, while taking a stabbing pass at the trunk, to effectively carve a small section of it away. Conversely, if the stab was too far in toward the center of the tree, it would abruptly stop the plow in its tracks, with a sudden jolt to the operator, as the greater density of the wood immediately forced a necessary adjustment to the stinger's trajectory, in order to more proficiently split and carve out sections of the trunk. This occurred every so often, while cutting into these particular hardwoods, with their inner density being so well noted, that quick adjustments had to be made, to notch out smaller chunks per each pass, until enough of the trunk was cut away, to eventually cause it to crack and fall.

In the potentially dangerous process of cutting a large tree down, we were always wary of its occasional tendency to twist away from the blade's control, to errantly topple back onto the plow, where it might present a serious problem, if not injury or death.

Quickly backing the tractor to one side, as the tree would begin its fall, while keeping the machine in close to the base of the trunk, usually ensured a clean getaway from it; while on the other hand, if the operator instead elected to move outward and away from the trunk, during that brief moment of the fall, it only invited the potential for disaster, as it was often next to impossible to look up and see just how these big trees were actually falling, while trying to back away. Also, if the tree were to suddenly splinter and twist, as they sometimes did, it would likely change the trajectory of the fall, making its projected landing area quite uncertain. So, the main idea, in order to remain in control and out of harm's way, was to stay in close, as the tree began its fall, in positioning the tractor just to the immediate side or rear of the trunk, where it was mostly removed from the falling tree's apparent line of descent.

At times, as if in defiance, these notched-out, splintered trunks would actually twist and turn on their partially severed base, with enough force to break off, and astonishingly, walk right off the edge of it, while dropping down almost vertically, onto the ground, before finally falling over to lay prone upon the floor of the jungle. (*This actually occurred once, as I watched in utter disbelief*) At that point, the operator simply couldn't know which direction it might go, while quick responses were called for in these more critical situations. Influencing the trajectory of the fall with a solid push of the blade sometimes helped in that regard, but not always, as some of the more top-heavy and taller varieties of trees would occasionally just go their own way, despite all efforts to dissuade them from it. In the event where an inexperienced plow operator elected to back away from the tree during a fall, instead of staying in close to the trunk, the falling tree then had the potential to impact upon the tractor, with its significant weight and downward force being more likely to damage it, even to the point of severely crushing the cab, if the two happened to come in direct line during that crucial moment.

This also happened on occasion, mainly due to inexperience, with an operator not fully attending to the details of tree cutting, and not reacting in a timely fashion. Consequently, a few injuries occurred every now and then, as a result of these unfortunate accidents. But luckily, no fatalities from these mishaps were ever recorded within our unit.

The steel cabs on these plows offered a great deal of extra protection to the operator for this kind of work, but were really no match for a thick and heavy, falling hardwood tree, to suddenly impact upon it with a direct hit. This is why only experienced operators were allowed to cut down the big trees, with the subsequent use of C-4 plastic explosives also brought into play, in bringing down the unusually large and more difficult ones later, by other combat engineers, who regularly handled the otherwise stable, yet moldable, clay-like substance.

Another large native varietal tree, that we occasionally encountered, within these tropical jungles of South Vietnam, was the Hopea (*pronounced, Hopay*), which featured tall, leafy canopies, and had wide, wing-type, buttressed trunks, that provided the trees with more support, for their large girth and height. But, these winged buttresses made it somewhat difficult to get in close, in order to cut into the main body of the trunk, as they occupied the lower outside portion of the tree, and regularly fanned out from the sides of the trunk, with their fin-like appearance running down to join with the tree's root structure below. Encountering these, as the unusual winged sections added to the trunk's overall thickness, it made for a more tedious process, by which to carve away those annoying protrusions first, in finally getting to the main part of the trunk, to more effectively slice through much of it, whereby, it would eventually bring the tree down. But, because of that particular difficulty, and time constraints, oftentimes these large buttressed trees would, again, simply be left in place for later demolition, using plastic explosives.

At the time, we didn't actually know the proper varietal name for this unusual tropical tree, with its rather peculiar built-in support structure. It was just a large, odd-looking tree, with these strange winged-type fins that made them look much like Banyan trees. *I later researched it, matching its overall description and locale, to more closely identify it as the Hopea tree.*

While working in the Iron Triangle area, as I steadily steered my tractor along in the cut, I was trailing a fair distance from another plow, and was unaware that a friend and fellow operator ahead, was cutting a medium sized tree, at the time, just up around the bend, and out of my immediate view. When I finally came crawling around the turn, I noticed the tree that he was working on had already fallen; and as it had leaned away from his control, it errantly fell over onto a nearby APC, striking the security's .50 gunner, whereby it necessitated an immediate call for a Med-Evac chopper to 'dust off' the accidental casualty to the nearest field medical hospital, as this man was suddenly in critical condition.

In the wake of the accident, my friend was thoroughly disconsolate that night, as he came to grips with the severity of what had occurred, while openly blaming himself for this horrible mistake that weighed so heavily on his heart and mind. In hopes of possibly raising both of our spirits, and setting our worried minds to rest, I then suggested that we walk over to where that particular APC was parked against the perimeter berm, and inquire as to the current condition of this '.50 gunner'. But, as we approached their imbedded position, their guys were somewhat hostile toward us, while tersely asking what we wanted, as they were clearly agitated and upset by the day's unfortunate event. After posing our question, while their COM radio squelched intermittently in the background, one of them curtly replied, "Oh, he's dead…ok? Now, get the hell out of here, and stay in your own area!"

Sauntering dejectedly back over to our squad tent, we were then left with silence between us, as we both struggled in disbelief, with the sudden reality of what had actually resulted from the earlier accident. As that fateful day finally came to a close, I saw that this tragic news further devastated my friend, and he quickly sank into a deeper funk and became totally inconsolable, as indeed the security's '.50' gunner had died from his injuries. In the days that followed, the deep depression that developed in him, wound up affecting his entire involvement with Land Clearing operations, by his own undoing, as he refused to get back on a plow again, and was soon sent back to Long Binh, where he remained until rotating out, psychologically impaired and scarred forever by the outcome of this terrible accident. In his defense, I can't say that it was entirely his fault, while that APC was actually positioned fairly close to the cut, in a spot that it shouldn't have been in, indicating that these members of the security weren't fully paying attention, as they were apparently unaware of the potential for this particular kind of overhead hazard. They simply took it for granted, that we would watch out for them, as they watched out for us, and it all went so terribly wrong. While it was painfully clear to them that they shouldn't have been positioned that close to the edge of the woodline, the security made certain that it would never occur again, as we all tried to resume with things, and get on with the mission, as best as possible.

Still, I realized that a conscientious tree cutter would have likely become more aware of things prior to felling that tree, even to the extent of waving the APC out of the immediate area before bringing it down, as safety was always the main ingredient in tree cutting (*it had to be*). Cutting trees with a Rome Plow was more of a sensory sort of undertaking, as it was a skill that required a more heightened level of ability, along with a keener awareness of all the elements involved with it, in order to be well accomplished in this systematic task. It wasn't an endeavor to be taken lightly; otherwise, the danger factor was great, as there was often little or no room for error. This is where inexperience sometimes created accidents, as most of our operators felt experienced enough to tackle tree cutting, but many were not; and occasionally, as in this case, (*with tragic results*) it showed. Unfortunately, that's the stark, bare-bone reality of it, and somehow I think our friend, Shepard would have more than likely agreed.

I refer to these unfortunate events as *accidents*, simply for lack of a better word. But, I feel that some of them might have been prevented if newer plow operators had been subjected to a brief, one-week training program with a more experienced operator, much like that which I had benefited by, when first coming to Land Clearing. As it was, though, operators at that time mainly learned from following the rest of the pack around the trace, and often discovered many things on their own, by trial and error, as they were put on the equipment and expected to perform, after acquiring only the basic operational knowledge of the machine itself. In time, though, most everyone adapted well enough, and eventually found their place in the cut, while a few others, who hadn't trained on heavy equipment, struggled, and were more prone to making mistakes; some being minor ones, while a few others were more critical.

With the addition of more Land Clearing companies in the I and II Corp tactical zones, as the Rome Plow tractor and its unique capabilities had gained more notoriety outside of III Corp, trained 'Crawler Tractor Operators' (62E20) were, at that time, becoming somewhat more scarce and less available to us, in replacing those who were either rotating out, or those few others who had been killed or severely injured. In those instances, the company was reluctantly obliged to fill the occasional vacancies with a few inexperienced infantrymen, or artillerymen, as they were the only replacements that were available to us, at times. This tended to contribute a little more to the problems that occurred, due to the overall lack of experience and in-depth understanding of the heavy equipment. A few of these new replacements, who were previously trained for something other than heavy equipment operation, turned out to be somewhat inept and not very mechanically inclined, while even showing some signs of tentativeness on the plow.

While we were all well acquainted with the fairly easy task of loading our tractors onto the lowboy trailers, whenever called on to do so, I remember an incident where one of the newer untrained replacements had a difficult time of it, as he somehow managed to position his plow on the trailer with part of one track hanging dangerously over the edge of the flat-bed, where he couldn't correct it, no matter how vehemently the squad leader lit into him, to properly right it, as he was terribly afraid that it would slip off of the trailer if moved any more than it was. As it turned out, this man somehow became extremely fearful of the situation, and simply froze, as I could see the sheer panic that had come over him, while I stood there nearby. It was mainly due to what I perceived as unnecessary stress from this otherwise, relatively easy task, and it was somewhat hard for me to fathom, as to why something so simple would cause him to 'freak out' like that. But, as he was untrained, and only knew the basics of how to operate these machines, uncertainty and doubt had apparently gotten the best of him, as he had fallen prey to his own unwarranted fears, in trying to do what he could, to right the tractor. With our squad leader angrily dismissing him from it, he then turned to me, and ordered me to right it, as if it were just a mere routine, everyday task, which I had also naturally assumed that it was. So, in slowly and cautiously moving the skewed tractor from its extremely precarious position, there on the edge of the trailer, the left track suddenly slipped from its partial grip, as it lost its bite from the trailer's edge, and slid off sideways, to leave the plow angled diagonally there, with one track still left up on the load bed, while the other then rested on the ground below. I believed that I had made a good try at righting it, but it was just too precariously set, where it was, to safely maneuver it back, and properly center it on its carrier. Just as that occurred, I could hear the expected blaring expletives, spewing forth from our squad leader, as he was absolutely livid with the outcome. Suddenly, it hadn't seemed quite as easy as I had originally thought, with the resulting mess leaving my ego a little bruised and embarrassed, even though it was somewhat of a strange fluke. Consequently, the inexperienced operator was later relieved from all equipment operation, and was sent back to our battalion area, where our company heads simply got rid of him, as he was found to be of no further use to our cause.

One of our later casualties, 21-year-old PFC Paul Quaglieri, of Burbank, CA, came to us in January of 1969, as an 11 bravo (*infantryman*). He had only brief initial training on the plow before being sent out into the cut; and while working in very hilly terrain around the Bien Hoa area, in March, his plow slipped off of a ridge and overturned several times, as it plummeted to the bottom of a steep ravine. There were no seat belts in the plow to restrain him, as none of us ever had them. His twisted, lifeless body was later found lying within the separated and mangled remnants of the plow's steel cab, where it had finally settled there, down amongst the wreckage.

<u>Paul V. Quaglieri</u> Age 21, from Burbank, CA

Date of casualty: 13 March 1969

Panel 29 W ~ Line 32

It had later been my own personal observation, in hindsight, that it was probably a mistake to have assigned him to a plow so early, without instructing him further on the cautionary operation of this machinery in hazardous terrain. He simply didn't know enough, as he wasn't trained on bulldozers, like most of the rest of us, and didn't fully know what to expect in certain situations, or while working in the more extreme landscapes. With the tragic loss of Paul Quaglieri, it was at that point, much more evident, that the shortage of trained bulldozer operators had then, sadly, come back around, full circle, to haunt us.

On one overly adventurous day, while again cutting in dense, high canopy type woodland jungle, with large, overgrown teak trees present, I felt like I needed to test my mettle, and exercise my own ongoing tree-cutting prowess. With the surrounding heavy brush and smaller trees already cleared away by the other plows in the cut, a few very large teak trees remained standing, looking fairly ominous, as they loomed over the newly cleared expanse that was formerly part of the contiguous jungle, which had fully encompassed the area only a few hours before. On some of the earlier operations, I had brought down a few notably large trees, but the ones encountered on this day were even larger, compared to those others. After selecting a considerably sizeable one for my challenge, I pulled my tractor over to a tree shaded area, away from the cut, where our Deuzenhalf was parked along the roadway, and got down and asked my platoon sergeant if it might be ok to "level that big tree across the way"? He didn't really like the idea, and advised me not to, suggesting instead, that we might leave the big trees for other engineers to blow up later, with C-4.

With that, he had essentially given me an indirect order to desist. But, in my own bullheaded sort of way, my desire to cut down that big tree somehow seemed to overshadow that, and I proceeded to do it anyway, while brazenly acting on my own authority. As I recall, it took me several passes with the 'stinger', while repeatedly stabbing it into the trunk, as the corner section of the blade carved out large chunks of wood in the process; and with each pass, I maneuvered the tractor to pierce it again from another angle, until it finally began to crack and then lean. At that point, I backed the plow around the tree, while remaining in close to the splintered trunk, just to an immediate position where the imminently falling tree wasn't headed. But, as it began to further crack and break, from the blade's telling effects, it unexpectedly twisted and rolled back toward me, while breaking quickly, and falling fast, as it left me but precious time to back away further. As I pulled back, it came crashing down with tremendous force, while the body of the trunk glanced off of the top of my blade, as it sheared off a section of the brush guards, and gave me and the plow a blow that slammed the elevated blade downward, while consequently bouncing the rear of the tractor up into the air, where it shot me up and out of my seat momentarily, to where I nearly banged my head on the inside top of the cab, before falling back down into the operator's seat, as the rear of the heavy plow dropped back down onto its tracks, with its somewhat rapid return to earth.

With the passing of that amazingly violent moment, as the dust from it soon cleared, it was plain to see that the large fallen tree had finally come to a horizontal rest, after toppling onto what was formerly the forest floor, with its monstrous trunk and canopy spread out over a distance, and with its outstretched branches covering the nearby roadway. Almost immediately, I whirled around, and met the prone trunk about midway, as I gathered it up against the blade, and pushed it back out into the open area, thereby clearing it out of the road in one sweeping motion, just as our Company Commander drove up in a jeep, with our security's Division Colonel riding along. I then went back over to where the remaining splintered and severed trunk base was, to gather up the few broken pieces of steel brush guards that would later be re-welded back into a fairly close facsimile of its original configuration, atop the K/G blade.

At the end of the day, after pulling maintenance on my plow, I learned that our CO (Captain MacNeil) wanted to see me; and as I made my way over to where his COM trailer was, I had thoughts of possibly being reprimanded for disregarding the Sarge's orders, in my obsessive determination to cut down that big tree. Instead, he sat me down and handed me a cold can of beer (*a luxury that we weren't accustomed to in the field*), and complimented me on the "fine job," in handling the earlier tree-cutting task, as the Division Colonel was greatly impressed by it. Upon hearing that, I was simply flabbergasted at that point, and just sat back and enjoyed my beer. (*What the heck!*)

~ Fourteen ~

Up in Smoke

With the daily passage of time, it was occasionally interesting to note that sunrises and sunsets in Vietnam were often more spectacular than the ones I recalled from back home in California, where I would sometimes gather with friends at a roadside turn-out, on an east bay mountain side, to drink a little beer and watch the sun slowly sink below the far off span of the Golden Gate Bridge, as the renowned structure occasionally became silhouetted against the backdrop of a soft maroon colored light. But, somehow, the different hues of bright orange and red seemed more vivid in the 'painted skies' of Vietnam, as clouds would marry with the accumulated smoke from the daily explosions, gunfire, and other war related emissions, making it seem like we may have actually had a hand in forming our own impressive morning and evening sky shows.

Prior to joining the Army, I had smoked a little pot (*marijuana*) back home, recreationally, with school buddies, and thought that to be the most potent stuff around, until I tried the 'ready rolled' stuff the hooch girls would provide, in re-packaged cigarette packs and cartons, back at the battalion area. More and more guys were smoking pot over there, as a drug subculture had established itself, to some extent, drawing ire and some unrest among those who instead chose alcohol as their main form of relaxation and abuse. For me, it was a way to relax at night, and while a few others in the company advanced to 'speed' or 'liquid opium', I just smoked a little pot from time to time. Although, on one unusual evening, when walking down to the EM club, in Long Binh, I encountered an old 'Papa San', who was working around the trash bins and dumpsters there. As he motioned for me to come over, with a friendly underhand type gesture, I curiously complied. When I got up to him, he presented a pipe that he was smoking, and thinking that it was just pot, I took a big drag and handed it back, while thanking him with a smiling nod, and continued on into the club. About the only thing I remember in there, was ordering a beer. After that, I was suddenly wiped out, as a few others there quickly rescued me and took me back to the billets. Apparently, that pipe that I smoked, had opium in it as well, and it was no wonder that it wiped me out. After that, I steered clear of other people's stuff, and just occasionally smoked my own.

Looking back, though, it's not something I'm proud of in any way, as I'm totally opposed to that sort of involvement now. I just got caught up in the subculture of the times, as many other 19 or 20 year olds did then, and I regret that it alienated me somewhat, from some of the other guys. Fortunately, as I don't have an addictive type nature, neither physically nor psychologically, I was able to eventually wake up and push myself away from it, while returning to a more sensible drug-free approach to things. One good thing is that I was able to fully realize the psychological effects that come with its usage, while witnessing that it tended to alter one's personality and affect their motivation, character, and overall outlook on life. For those who may not view marijuana as a potentially dangerous drug, they definitely need to look again, as it changes people's nature, and tends to subtly remove them (*psychologically*) from their true self.

Unfortunately, at that time, drug usage had become more of a serious problem among the troops in Vietnam, as its availability was greatly fed by the local post workers, and it made a serious contribution toward lowering the overall level of professionalism within the ranks. With the anti-war movement progressing back in the states, and our government's winning of the war left in question, with little or no sense of urgency, it was a little easier for some of us to give in to temptation, and try some of the items that the hooch girls and young children along our convoy routes made available to us.

Involving a different sort of (*invisible*) substance over there, my sudden exposure to CS gas became a rude and surprising encounter one day, while cutting through partially defoliated jungle near Tay Ninh. When I first noticed the operator ahead of me, as I approached his stopped plow from behind, oddly, he was standing on top of his winch drum, while vomiting and acting strangely anxious and disoriented, as if something distressing had just occurred. But, I couldn't figure just what it might be, as there was no sign of enemy presence about, and he seemed to be distressed in an entirely different way.

Then, as I stopped my plow just short of him, something in the air unexpectedly overcame me, and suddenly all I wanted to do was, to just jump out of the cab and flee the immediate area, which I did, with reckless abandon. I managed to get a little distance away, and as the air was clearer there, I recovered enough to eventually make it back over to my tractor, and quickly backed it out of that spot, with the other operator coming onboard with me. In this little incident's aftermath, it was discovered that the other operator's blade had punctured a 55-gallon drum of CS gas that the Air Force had dropped previously, intending for it to burst on impact. However, with this particular drum, things didn't quite go according to their drop plan. The telling effects from this chemical agent had my eyes and nasal cavities burning tremendously, while it left me coughing violently and gasping for fresh air.

I remember initially being exposed to CS gas in basic training, at Fort Lewis, along with tear gas, as part of a training exercise to get us more acquainted with these chemical agents, in learning how to react when they suddenly become present. The CS certainly got everyone's attention, and I never could see how anyone could actually do anything while in its presence, except to flee the affected area. CS isn't particularly harmful, that I know of, but it is extremely disorienting, while being quite anxiety producing, as it immediately alienates any and all who come in contact with it, making it next to impossible to function normally under its influence. Incidentally, that particular afore-mentioned sector of partially defoliated jungle actually turned out to be attributed to the Air Force's additional usage of the defoliant 2-4-D, or *Agent Orange*. Essentially, we were sent into that area to finish a job that the 'flyboys' weren't able to fully accomplish with their toxic airborne chemicals.

In yet another unusual incident, near Tay Ninh, while I was cutting along the backside of the established trace, on one occasion, late in the day, one of my outer hydraulic lines suddenly broke at the upper housing on the left push arm, as it instantly spewed hot oil out to the side, and fortunately away from the interior of the cab. In response, I immediately shut the plow down and got out to assess things, where I soon found that it was rendered completely inoperable, with the heavy-duty coiled-steel and rubber hydraulic line being irreparably breached the way that it was. Whenever this occurred, it was always best to just shut the engine off and let everything cool down, in advance of our maintenance crew's arrival onto the scene. Otherwise, as the engine continues to run, without pressurized oil within the hydraulic system, essential inner seals tended to quickly burn up or melt, with the excessive heat that would soon follow, necessitating a complete overhaul of the system. So, all I could really do at that point was to sit it out and wait for our security and maintenance boys to come around and find me. Although, the way the plow was positioned there, it was more inside the cut, being mostly obscured from view; and as I thought of throwing out a smoke grenade to establish a 'visual' on me, I noticed that I was fresh out of them. So, I just sat there and waited, still feeling secure that they would soon come around to the backside of the cut and discover my 'downed' location.

Within minutes, after sitting there in restful silence, I stirred and fidgeted a bit, as the stillness within the jungle was just a little unsettling; but naturally, I decided to continue waiting there until hearing the familiar sounds of our machinery approaching, or until catching sight of any of the other plows or track vehicles. But, it seemed strange that I heard nothing, not even the tiny din of a motor in the distance; and as the sun eventually went down, and darkness began to envelop the area, I realized that something was definitely wrong. As I continued to sit there in the dozer's seat, I grew more leery of the possibility that the enemy might instead, be the first to find me; and, in gathering up my rifle from its place behind the seat, I quietly 'locked and loaded' a full magazine and exited the cab to take up a new position, down between the radiator and the back side of the blade; which afforded more protection and room for firing my M-16, in the event of a 'worst case scenario'. I thought that if enemy elements did approach, they would likely try to get into the cab first, in search of an occupant; and with my lower position there, I could at least defend myself from a better vantage point, if they somehow happened to spot me sitting there, in the extreme darkness that the jungle cover had further provided at night. Locked into a more heightened state of awareness at this point, while growing ever more anxious, I was completely taken aback with my sudden plight, while fully realizing that this was indeed becoming a real situation; one in which I wasn't sure that I was fully prepared for. However, I reassured myself, that I was ready for whatever might come.

At that point, with a medium growth tree canopy directly overhead, along with a few noticeably luminous stars shining through from the moonless starry sky above, it was getting close to pitch dark, there within the edge of the woodline, and the deafening silence ensued. It had never really occurred to me that this strange sort of situation could actually happen, as I further came to grips with the stark reality that somehow I had been left behind to suddenly fend for myself. What might have originally seemed more like a normal mechanical breakdown, had then manifested itself into a full-blown scary predicament, with the simple turn of day into night, along with the unusual absence of my fellow engineers and our mechanized security elements. Eerily, there was no one within earshot, and no one in sight. To say the least, my adrenalin was up.

Sitting there, with my back against the rear of the K/G blade, and my rifle at the ready, I was unsure whether I should leave the plow and half-blindly walk down to the corner of the cut and make my way out across the long debris-laden, open expanse, in order to reach the road, and walk the mile and a half or so distance from there in the dark, on down to our NDP. But, too many doubts arose, as I played out that possible strategy in my head, in hopes of alleviating the overall situation. I reasoned that if I were to grope and stumble my way out across the field in the dark, and somehow manage to make it all the way back around to the other side of the cut and over to the road, there still wasn't full certainty that I might succeed in getting down that road and back into our NDP. Plus, I then came to the further realization that, if I just happened to make it all that way, and still went unnoticed by the enemy, once I got within range of our NDP, I might still be mistaken for *'Charlie'*, and erroneously draw friendly fire from our own security force. So ultimately, I decided it would be best to just remain with the plow and continue to take my chances there, for the time being. It was simply too dark, with too much obstacle-laden terrain lying between my plow and the far off roadway, and there were just too many other doubtful questions hanging over my head, as this stressful and perplexing problem wore on.

Holding my silent position there, I still couldn't quite figure out what the problem was, and why they left me behind. It just didn't make any sense, as it had never happened to anyone before. Late in the day, at the appropriate time, the plows were always rounded up from the cut and escorted out alongside the main access road, to head back down to our NDP for the night, just like normal clockwork, with this particular day unexpectedly becoming the lone exception.

In the presence of this weird silence within the jungle, it seemed so odd that there weren't even the chirping sounds of birds in the distance, nor the buzzing noises that certain insects give off. This was the jungle...and it sounded so uncommonly dead, as if my being there had actually influenced its muted effect in some way. The entire area was simply devoid of sound, and it felt downright creepy, as I strained to hear nothing at all, except for an occasional soft metallic creak from the tractor, or the light whisper of leaves rustling in the breeze. The spooky effect just added to my overall state of apprehension, as I remained resolutely silent and steadfastly in place. Of course, I reassured myself that it was likely the result of our plow's earlier active presence there, in the afternoon hours, which probably drove out all of the animals and silenced the insects. As it was, I wasn't really sure, and was probably growing a little delusional in my current state of anxiety. But, I just didn't want to think that it might otherwise be due to those other human-type predatory animals (VC) who may have been lurking about.

Due to the extreme potentiality of the situation, I had every reason to be nervous and fearful, as I knew full well, that all areas outside of our protected NDP were questionable and possibly suspect, with more heightened levels of enemy activity regularly occurring at night. Within the evening's shrouded cover of darkness, *Charlie* was much freer to move about, as they would oftentimes harass our NDPs at night with their usual intermittent rocket and mortar barrages, to continually remind us of their constant opposing presence.

As I continued to sit there in silence and dreaded anticipation, for what seemed to be more like an eternity, my mind began to race in every direction, as the uncontrollable flow of adrenalin made me feel more alert and aware than ever before. I was neither hungry nor thirsty, and far removed from any thought of either one. However, I did crave a cigarette, but knew all too well how utterly stupid that would be at a time like this. At that point, I just didn't know what else to do, in seeking to remedy things, as I desperately wanted an end to come to this quite hairy and helpless situation.

Then, as if heaven sent, I suddenly picked up the distant sound of engines and noticed a dim flash of light coming from afar. As the familiar sound and lights got closer, I left the confines of the plow and ran out into the clearing to flag them down. Their vehicle lights shone brightly upon much of the downed debris that was scattered about in the field. It was a couple of APCs from our security, along with our M-548 maintenance track… and man, was I ever glad to see them! Our maintenance guys brought out a tow rig and pulled my plow back in to our NDP; and as I ran into my platoon sergeant upon arrival there, he was somewhat apologetic, as he explained that the plows had been mis-counted prior to leaving the cut, and no one had suspected anything until someone finally noticed my machine's absence during motorpool maintenance.

I imagine somebody must have had to answer to the CO for that error, and more than likely got chewed out as a result.

For me this was an ordeal of extreme uncertainty, out there alone in *Charlie's* element, where he generally ruled the night, as I was stranded without any support or line of communication, for a little over two hours; and although I was pleased and relieved to have finally been rescued, I wasn't entirely a 'happy camper' back at the NDP. It clearly angered me, that this sort of thing had actually happened, and that a simple mathematical error might have otherwise gotten me killed. I was all too relieved, though, that things had turned out the way they did, and that the enemy apparently hadn't detected my presence there, if they had even been anywhere near the area that night.

In retrospect, I think that perhaps I was just extremely lucky to have gotten away with it on that particular night. My resolute silence and inactivity there could have quite possibly been my only saving grace. In that regard, I knew full well that sound tended to travel farther during the evening hours in and around the voided areas of the jungle. If the enemy had actually been out there and discovered my presence, I would more than likely have been outnumbered and overwhelmed with small arms fire, to the point of being taken out of the picture (*or captured*). Consequently, the plow would have possibly been rigged up with trip-wires and explosive charges, to serve as a greeting card for those who would eventually come looking for me.

~ Fifteen ~

Redemption

At some point later in my tour, I noticed that the majority of our winch cables had fallen into serious neglect, as many of them had become frayed, and some had even lost their heavy steel clevis and hook, which were normally attached to the looped end of the cable. So, I voiced my concern to our Platoon sergeant that each of these were in various states of dis-repair, and that they were in severe need of attention if we intended to use them at all. Without a proper cable eye-loop for the hook, many of the winches on these plows were rendered next to useless, even though they still operated perfectly well, with the unused spools of steel cable still remaining tightly wound up on each drum, while their frayed ends lapped over and partly hung down, to look rather absurdly like a cow's tail. A few of the plows were even missing the entire spool from the drum, leaving those operators without a means to pull their tractors, or any of the others, for that matter, out of a disabling situation. In pointing this out, I had also mentioned that I had prior experience in re-weaving and clamping the winch cable eye-loop, so as to restore it to its original intended configuration, by effectively repairing this seldom used, but needed component on the plow.

In my earlier days back in Bearcat, with the young 86th LCT, we had all learned how to properly re-weave the heavy, multi-stranded cable, to duplicate the factory woven eye-loop, which carried the locking clevis device that secured the large winch hook to it. In fully restoring these, a correctly woven eye-loop greatly improved the pulling strength, and increased the winch cable's overall reliability, as it was sometimes utilized and tested in a few unusual and more challenging situations.

Since I was one of only two men remaining from those days, and was the only one who knew this procedure, it was put to me to conduct a type of night workshop in order to teach this to the other operators in an effort to get all of the plow's winches fully operational and back in original order. So, I performed the 'weave' in front of the other guys; and along with sergeant Watson and a few of our mechanics, I went around to each plow and bull blade afterward, to ensure that they were all getting it right, in creating a small, tight, interlocking loop for the clevis and hook, which was re-woven in such a way as to make it an integrated part of the cable itself. Supplementary to that, new spools of cable were added to the remaining tractors with empty drums showing, and before long, with the help of our maintenance team, all of the earlier sad fleet of Hyster winches were fully restored to working condition.

Much to the annoyance of the other operators, this little workshop lasted well into the night, and bit into our 'dream time'. However, I'm sure, when everyone realized on the following day, that they actually had a proper working winch, with a fully restored cable eye-loop and hook, they quickly got over the fact that they may have missed a little sleep, in accomplishing this necessary maintenance task. Shortly after that, SSG.Watson recommended me for a promotion, and I made SP/5 (*Specialist, Fifth Class*), which equaled the same pay grade as buck sergeant.

In the Army's 'subdued' type rank insignia, SP/5 (Spec-5) was represented by the SP/4 patch, (the O.D. green shield and black eagle), with an added black rocker riding over the top of it, to signify 5th class. In other parts of the world, away from combat zones, SP/5 patches sported a dark green shield, with a yellow spread-eagle and a yellow rocker above.

After receiving word of my promotion, a somewhat unfortunate incident occurred, just as we began one of our regularly scheduled Stand Downs on Long Binh Post. While settling back into our billets that night, one of the other operators had become openly annoyed and upset with this latest round of promotions, as he was overlooked this time, for the elevation in rank to SP/5 (*which was something I really had no control of*). But, through his angry outburst, he thought, since he had more time in grade, he should have gotten the promotion instead of me. For that matter, during the course of his pathetic ranting and raving, he even tried to provoke a fight with me over this, before the other guys stepped in to break it up. Looking back on that, I know now that the promotion had everything to do with my overall experience and expertise in helping to repair all of those winch cables, as well as my performance on the plow and extended time with Land Clearing. After all, at that point, I was the 'old guy' on the plow, and just naturally knew more than others, as a result of being a part of the 'team' for such a long period of time. Plus, it may have even shown that I cared about the equipment and the overall success of our field operations. 'Time in Grade', which signified seniority in rank, didn't always play into promotions (*especially in combat zones*), as this other man should have more fully realized before going 'ballistic' and making a complete fool of himself in front of the other men. In his emotionally irrational state, while trying to show me up, he instead, only succeeded in weakening his own stock with the other men around him.

Granted, I had known Sergeant Watson from way back, when he and I were both early plow operators within the old 86th's version of the outfit. At that time, Bobby Watson was a Spec-4, while I was a PFC, and he and I were fairly close friends, where at some point, we even occupied a squad tent together. But, over time, he achieved his rocker above the spread eagle (*Spec-5*), and soon after, had it switched to three stripes (*Buck Sergeant*), while becoming a squad leader, whereby a change had then come over him, and he was no longer the same gregarious squad-mate whose company I had so often enjoyed. As time then progressed, and as our previous Platoon sergeant rotated out, he was then given his lower rocker, becoming an E-6 Staff Sergeant, while meritoriously inheriting charge of 3rd Platoon. At that point, he had become even more removed from our former friendship, while I was mostly regarded as just one of the other operators in the platoon. Upon later learning that he had recommended me for promotion, it was just as surprising to me, as to most anyone else there, since it wasn't something that I had anticipated, and I knew that it hadn't been a product of our earlier friendship.

About a month after earning my specialist's rocker, an opening for squad leader suddenly presented itself, as the previous squad leader had just rotated out to the states. It wasn't anything that I was even remotely interested in, as I figured to continue whittling away my time, while locked into the daily rhythm and routine that went along with operating a Rome Plow. However, somewhat reluctantly, through the urgings of Platoon Sergeant Watson, I soon relented to the offer, and took on a much different task, as I filled out my remaining time within the 501st, in 1st Platoon. It meant that I would no longer operate a Rome Plow, and instead would ride the edges of the cut, atop one of our security's Armored Personnel Carriers, to watch over the clearing process in my appointed sector, and communicate any problems or situations through the use of a PRC-25 radio. It didn't provide any higher pay, or another jump in rank, just more responsibility, meaning that I had a squad of men and plows to look after. So, with that inevitable change, I soon realized that, as I had once followed others, accordingly, it was then time for me to help lead.

Here, I was checking the plows on our Lowboy trailers, soon after appointed as squad leader, to ensure that they were properly secured on the load beds. A local village boy stood by, to take in the somewhat amazing sight of our lengthy convoy, as well as to marvel at our impressive collection of specialized heavy equipment. July, 1969

In my new role as squad leader, each day I would sit atop a designated APC, upon a cushion, just behind the driver, who was at the extreme left front, positioned inside, and down in a hole, where only his radio-helmeted head protruded to provide for the view. One of the other men aboard this track vehicle had an M-60 machine gun on a tripod, behind a metal protective shield, with sandbags built up around it, while another was equipped with an M-79 grenade launcher and another two or three had their M-16 rifles (*occasionally equipped with high-powered scopes*). The '50' gunner, who also commanded this fighting vehicle, was positioned in the center, and forward of the others onboard. He always wore a radio headset, with a microphone, for direct communication with the driver, and with all of the other mechanized infantry vehicles in the area, which generally comprised our field operation's entire security force. He was situated just to my right hand side, and sat behind a protective metal plate, or shield, in the form of a semi-wrap-around shroud of armor plating, that swiveled with the big heavy-barreled machine gun that he controlled.

In supplying the assorted weaponry onboard, several steel ammo cans were stored on the deck, containing brass-jacketed 7.62mm rounds, along with the larger .50 cal. rounds, which were all on bandolier type clips, for continuous feed. Also, metal jacketed, 40 mm fragmentation grenade rounds were present, which looked like big fat bullets, and were fired from the shotgun-like M-79 grenade launcher, with its short single barrel, that broke open at the breach for re-loading. This handy, light-weight weapon was quite similar in style to the old double-barrel shotguns from earlier times; except that they fired actual fragmentation grenade rounds instead of buckshot. While these ammo cans were regularly kept up on the deck, more ammo was stockpiled below, as there was plenty of ammunition available to provide ample supply for these various weapons over an extended period of time, barring any lengthy firefights. With the occasional switch of security elements, I would ride atop a different APC from time to time, with an all-new cast of characters onboard. But, all were similarly equipped, and diligently dedicated, just as the previous group had been, as they alertly policed our concentrically cut-out work areas, while we labored to complete our appointed task.

Here's a plow operator's view of me, while I was seated atop an APC, directly behind the driver.

A good supply of M-18 smoke grenades, along with standard fragmentation type grenades, were also available to these men, as different colored smoke was often used for signaling to aircraft, or to others on the ground, and to provide cover or screening for troops. These cylindrically shaped grenades featured red, green, yellow or purple smoke, with each different color being mainly recognized and reported, via the radio, in relating or confirming, that a particular color of smoke had just been "popped", for whatever the intended purpose. As far as their use for signaling while aboard the plows, the colors really didn't matter all that much, from the standpoint of the operator. They were mainly noted in communicating a plow's downed location, while relaying the information over the radio, so that the security could acknowledge any potential enemy activity and quickly move in; or so that our M-548 maintenance track could gain a visual on a disabled plow's position. Otherwise, the only notable, yet curious benefit that a particular colored smoke may have presented to us, as we worked, rested with the amazing effectiveness of green smoke, in miraculously deterring swarms of angry bees.

Taking up my position atop the APC, I found a little additional space available for stowing my M-16 rifle, where it would regularly remain beside me, resting alongside the radio, in the possible event that it might be needed. While moving about, I would instruct the driver on where we needed to be, in staying within close proximity to the plows, as we often had to ride over logs and other downed debris, in moving closer to the ever diminishing woodline, until the final swath was cut, and the plows would subsequently follow the lead plow over to the next targeted area, where the entire process would be repeated. This regular cycle of activity would continue each day, until the order was given, late in the afternoon, to round up the plows and head on back to our NDP for the night.

While there were sometimes a few different variations of the aforementioned weaponry onboard our security's track vehicles, each APC remained highly armed and fully capable of defending most attacks or ambushes that would occur from time to time as we forged ahead with our mission, out around the cut. All the men up top wore flack jackets and most wore their steel pot helmets. They intently watched the shrinking woodline as the plows worked, for any sign of enemy presence that they might otherwise respond to. In the event where they found it necessary to dart off in pursuit of suspected activity, I would quickly gather up my rifle, radio, and my canteen, and would drop down over the side so as not to get in their way, as my presence wasn't needed in those particularly tense involvements. Typically, with those items at hand, I would just hold my position there, while awaiting their eventual return to pick me back up again. While temporarily stranded there on the ground, I would basically resume watch over the plows and continue to monitor the radio, while scanning the near vicinity for any sign of unusual activity. On these sudden breakaway occasions, I was then simply appointed by default to secure my own immediate surroundings for the time being.

These commonly used M-113A1 Armored Personnel Carriers were powered by a compact 6-cylinder Detroit diesel engine, and were mainly constructed of aircraft aluminum. This made them nearly as strong as steel, yet were light-weight for carrying several men, along with a good supply of ammunition and other personal gear. They had a reliable 3-speed automatic transmission, and each offered good power and maneuverability when traversing over different obstacles and terrains. Additionally, these versatile track vehicles could even swim across small bodies of water, if called for. Their standard lightweight metal shell allowed for more than adequate floatation, while the churning tracks provided marginal propulsion and steering. A lower, rear hatch folded down to provide a ramp for accessing the small, inner confines of this track vehicle, which doubled as the two-man crew's living quarters. The somewhat cramped interior of these rolling 'tin cans' featured a couple of long bench type seats that folded out from the walls to provide a gathering area at night. They also served as crude bunks, to give a little 'sack time' to the men, as they alternately stood watch and slept. Otherwise, the other attached infantrymen onboard caught naps however and wherever they could, depending on their location and overall situation. The aspect of getting a halfway comfortable, full night's rest never really played out well for our security people while they were in the field. Comfort was just not part of the assignment for any of them, at least up until such time when the operation would finally come to an end and they'd return to their main base camp for their own particular Stand Down period.

Like the others onboard, I wore a flack jacket some of the time. It supposedly helped somewhat in slowing down any potential shrapnel from land mines and RPGs. These sage-colored jackets were sleeveless and looked a lot like a vest, but were kind of bulky and noticeably heavy to wear, which caused some of the plow operators to disregard their use. I was often persuaded by the Sun to discard it as well on some of the region's more oppressively warmer days. It wasn't at all like a bulletproof vest, but it at least offered some marginal protection that could mean the difference between life and death, in some cases, as compared with having no protection at all. However, in the heat of the day, it tended to lend itself to unnecessary feelings of discomfort. Its chosen usage on those warmer days would often encourage one's sweat glands to increase their normal output. I also donned a padded fiberglass helmet that contained earphones and a microphone, which plugged into the AN/PRC-25 radio unit that I kept down beside me onboard the APC. Other times, a helmet-less set of earphones with an attached microphone often accompanied the radio.

The "Prick-25", as we knew it, was a little bulky and heavy to carry around, weighing in at about 24 lbs., with its sizeable battery pack. But, it was actually a rather reliable, portable FM (*high-frequency*) field transceiver that offered several hundred possible channel positions to the user (*920, to be exact*). It could clearly transmit and receive 2 to 3 watts of power over about a three to seven mile radius, depending on terrain. Aside from being simple to operate and relatively trouble-free, it was durable enough to withstand major impacts from high-level falls. Plus, it was even waterproof, minus the headset. Firmly established as the military's standard two-way radio in Vietnam, it was also commonly found mounted within vehicles and helicopters, as well as being largely utilized by ground infantry RTOs in the field (*Radio Telephone Operators*), carried mainly by backpack. In that respect, it was used almost universally in military application, over most other types of communication devices.

The radio, being our direct two-way line of communication, wasn't to be used for a lot of casual 'chit-chat'; only important operational communiqués, navigational assistance, and progress or damage reports, etc. In conversing, we regularly used the common CB (*citizen band*) lingo over the airwaves. The NATO phonetic alphabet was also used in referencing and describing certain things. To say the least, it was nothing like a phone conversation. While we repeatedly worked the tractors in cut, our security forces routinely monitored our COM channel, in addition to maintaining their own. They alertly listened to all of the chatter as they conscientiously stayed tuned in to everything that was occurring within the general area, in an ongoing effort to stay on top of things, security-wise.

As basic 'Standard Operational Procedure' (*SOP*) in the field, we religiously maintained radio contact throughout each workday. Moving about from place to place, as I kept pace with the plows, my platoon sergeant or platoon leader would regularly direct me and provide pertinent information for my assigned sector each day. Like clockwork, I would, in turn, routinely check in, with a progress report on the hour, to generally keep everyone apprised of our activities there. Just before departing the NDP each morning, I was given a topographical map for the particular areas we were slated to work in, which contained a grid that divided the map into various sectors. We used it to plot and find our way around through the application of an extremely accurate ground navigational system called, the *United States Army Military Grid Reference*. In general recognition of this reliable course-plotting method, we commonly referred to it as "Grid Coordinates". When applied and figured correctly, this common navigational system generally placed us within yards of the projected location on the map. Using a 'topo' map to figure and judge distances, according to this system, each little hash mark at the edge of the chart represented a kilometer, and was referred to as a 'click'. The distance figured from point A to point B, was found by counting the many clicks (*or kilometers*) at the edge of the map, which corresponded with the two established reference points.

While sitting in place, monitoring the movement of the plows along with the day's progress, we would often remain in one position for at least an hour or so. It tended to get a little boring from time to time, as some of these mechanized infantry guys would occasionally stir a bit and get a little cranky with each other during the course of the day. Of course, it's hard to relax and remain overly pleasant when you have to try to stay sharp and be on your toes, all the while knowing that *'Charlie'* might be lurking nearby, ready to initiate an ambush. I came to realize that some of these guys often went on little sleep, and were understandably irritable on occasion. Most of their time awake was essentially spent on guard duty, whether out around the cut by day or on perimeter guard by night, within our NDP.

Trying to adjust to the sudden role change as squad leader, I found myself feeling somewhat uneasy and bored with it at times. Sitting atop an APC that only moved as the jungle receded, while the plows routinely crawled around the trace knocking down trees and slicing through heavy foliage; I would simply make routine radio contact on the hour, to report on our status. Then I'd sit and continue to watch over things there, until it was time to move in closer to the cut, or until lunchtime came; or until word was given later in the day to round up the plows and head in for the night. Of course, this mundane activity repeated itself every single day. Needless to say, monotony was certainly a factor, with a lot of the intermittent associated boredom having to be endured along with it. However, this all changed a bit for our security, as their lull was occasionally broken whenever we encountered or suspected enemy contact of any kind around the area. These mechanized infantry elements would jump into action and always respond in a swift, but collective and cooperative business-like manner. In their aggressive approach, they tended to effectively root out the particular source of the problem, whether from the detected nearby presence of Viet Cong, or simply from an unseen animal whose half-hidden movement might have drawn a mistaken response.

Sitting atop the APC, contemplating my usual situation, I realized that I was then completely out in the open in an obviously vulnerable position, unlike the seemingly more secure, steel-enclosed confines of a Rome Plow. Without the reassuring heavy steel-plated protection available, which the plow and its screened cab had previously provided, I mostly relied on a group of relative strangers to compensate somewhat for that loss, in the event of a skirmish of any kind. Still, even though we were all fully exposed upon this armored personnel carrier, we knew we had the sufficient firepower to respond to most enemy incursions. However, in the event of an attack, there wasn't much in the way of protection for many of the men up top (*including me*), short of the gun shields that a couple of the gunners sat behind. Additionally, there wasn't any overhead cover to be found to effectively shade us from the intense rays of the sun on those exceptionally hot days that tended to invade and overwhelm our comfort zone. So, we just had to cover ourselves as best as possible with headgear and sun glasses, using our Coppertone sun tan cream to guard against burning. We also drank a lot of water while enduring the heat, and sweated it back out at a rather rapid pace. In fact, on some of the more oppressively hot days, the normal need to urinate was noticeably lessened, due to our sweat glands' overall level of hyperactivity. In the opposite extreme adverse conditions of Monsoon season, we had to rely on our plastic ponchos a great deal to keep us as dry as possible during the torrential cloudbursts that often ensued. But, despite the annoying conditions, we always continued on with our operations, employing nearly the same procedures, in monitoring the daily activity out around the cut.

Again, the day-to-day boredom was just something I had to deal with each day. It represented the same kind of lag in time that had more recently brought me back to Land Clearing, as I intended to keep the idle time going so I didn't have to think about it. As a result, I often wished that I hadn't taken on the duties of squad leader, as I would have still been out there on the plow, mowing down my time along with the trees and other heavy vegetation of the jungle. But, aside from that, it was the work with the machinery that I had missed; the hands-on repair work, and the camaraderie with some of the other guys that was now absent from my daily routine.

Although the other guys whom I had previously come to know, and had worked and interacted with for so long from those early days, had already gone back to the States, it still seemed so odd to me that some of their replacements had even rotated out after putting in their one year of service with the team. I had come to know them all; the earlier members and the latter ones, to some degree or another, and yet I still remained.

Suddenly, eighteen months was becoming a noticeably long tour, but I had weathered it fairly well over most of that time span, mainly using the Rome Plow as an effective time diverter. However, with my daily exposure atop the APC, I was beginning to get noticeably 'short' on my tour and, as a result, I was becoming more aware of my vulnerability from that position. It often bothered me and played on my mind during those somewhat dull, quiet moments while just sitting there out in the open, as if we were just waiting for something to happen.

Setting up an M-60 machine gun within one of our NDPs.

As a squad leader, it had become somewhat difficult at first to direct my guys and also be a friend at the same time. Of course, they always respected my rank and position, and followed any orders that I gave them; but our relationship had clearly changed with my newly assigned status. It was just an unusual sort of transition for me, at the time, which seemed to mirror the change that Bobby Watson went through, when he had earlier become a squad leader. I had been so accustomed to being part of the action among the others, and then quite suddenly had switched to become more of a director, spectator, and part-time babysitter. But, as with most adjustments, I soon came to realize that, with my status as squad leader, my peer group was to gradually change as well.

Here I am, looking all too serious, while sitting in one of my squad's plows, my left hand resting on the automatic transmission's gearbox. The long handle sticking straight up from my right arm is the 'stick' that controls the blade, the angled handle directly in front of me being the main throttle control lever. The two short levers that appear just below the throttle are the left and right steering clutches, and the two controls at my far left operate the rear drum winch. Also, positioned just behind the operator's seat is the diesel-filled fuel tank.

During Stand Downs back at Long Binh, I generally supervised the work in the motorpool for my squad and helped detail what was needed for our plows, along with the maintenance crew's usual involvement in getting them re-tooled and in shape for the next outing. Any 'hands-on' work that I performed then had to be done more in an instructive way, so as to pass any knowledge or repair techniques on to the newer operators. With my advancement in rank to SP/5, I was now considered an NCO (*Non Commissioned Officer*), and could no longer go to the EM (*Enlisted Men's*) club, but occasionally enjoyed the new surroundings of the NCO club, as I began to get a little more accustomed to the change.

As yet another field operation unfolded, we found ourselves once again back in heavy woodland-type jungle, among the large trees and their spreading canopies. Flying lemurs once again made their presence known, as they shrieked disapprovingly and protested our thundering presence beneath their otherwise tranquil, lofty domain. In continuing on, a few of the high-canopied trees were brought down just across from my position, where I sat and observed the nearby activity from my perch atop an APC. The violent downing of the trees shook up the flying lemurs in the neighborhood, and they responded with shrieks as they anxiously glided around from tree to tree. Strangely, one of them drifted down and landed on the ground; where it proceeded to awkwardly scamper over logs and other downed debris, its flabby excess of furry skin flopping around to create a somewhat uncoordinated looking gait. Moving along, it headed diagonally across the cleared expanse, somewhat in advance of my location. I then observed that three of my guys were directing their plows away from the cut to converge on the animal. They moved across the clearing to where the road was, pulling up just short of where the lemur was headed. Quickly exiting their tractors, they then ran up ahead to cut it off.

Seeing what was unfolding, I was angry that they would put themselves in harms way by trying to capture or closely observe this monkey-like animal. Plus, they were taking time away from the important job at hand. So, I tapped the APC driver's helmet and instructed him to get over there, fast, before one of the guys gets bit or clawed by this apparently agitated lemur. When we arrived, I immediately jumped down and ran toward where the three were waiting for the animal to converge on them, grabbing up a dead tree branch along the way. Just before I could get all the way up to where they were standing, the lemur scampered up within a couple of feet of them and stopped. Without further hesitation, I then quickly ran right through the middle of where they were standing, swinging the tree branch like a club, and struck the lemur square on the head with a couple of sharp blows, causing it to fall over onto its side, where it lay motionless.

The three operators standing there were shocked and upset with my sudden outburst, and asked me why I would do such a thing. They apparently judged the animal to be harmless and thought I was flipped-out crazy to do such a violent deed. I could only yell back at them to: "get the hell back on your plows, and don't pull another stunt like that ever again!" There may have been a few strong expletives interspersed here and there as I tore into them, but they just didn't see the danger with this type of animal and the potential for serious injury with any close encounter. I knew full well that these wild animals could be aggressive at times, with their sharp claws and fanged teeth, and that it was always best to just leave them alone. Seeing that my men had put themselves in possible danger, I felt that nothing could be left to chance, and was compelled to react to the situation as I perceived it. In the aftermath, I never did check on that lemur to see if it was dead or merely knocked unconscious, and I hoped that the two blows might have instead only stunned or injured it to some extent. It was just terribly unfortunate for the lemur that these guys didn't consider their own welfare, as their little escapade forced me to react in such a brutal way as to prevent the animal from attacking. Although the regrettable event had clearly bothered me from the standpoint of harming an otherwise innocent little animal, I always stood by my action, and never wavered from my reason for doing so.

Traveling out to the cut one morning as we skirted the edge of another woodline to get over to our targeted area, I smelled the foul odor of death in the air and caught sight of a rather grisly scene from my perch atop the APC. Just as we slowly turned a corner, I glanced down toward the right front side of the track and noticed this bluish-gray object that soon registered and came into focus as a man's mutilated upper torso, which was missing all of its extended components, just lying there on the ground like a big lump of meat, the unmistakable definition of its back and shoulders clearly in view. We surmised that it must have been a VC who had somehow been blown apart. But, no one could tell for sure, in the state of its decomposition and lack of anything identifiable, whether it was actually theirs or ours, so we just rolled away from it, assuming the former, and resumed with our mission for the day.

In my mind's eye, during some quieter moments of personal reflection, it had come to amaze me that I seemed to be getting more and more used to these encounters that were so closely associated with death. Prior to Vietnam, my only real close contact with the grim element of death had been among the domestic livestock and pets around our small ranch back home, as well as occasionally coming across the remains of an unlucky feral animal when wandering further out in some of the wilder areas of northern California. But, with the ugly specter of war being so prevalent there, occasional injury and death to others around me seemed somewhat more commonplace within this environment that had held me for so long, and was no longer all that shocking or unusual. I had simply grown accustomed to these extraordinary events and encounters over the long haul of my extended tour, and it was rapidly becoming more evident.

Many of our returning servicemen from Vietnam, as a result of their experiences there, came back a little traumatized and somewhat tormented, with the frightening presence of occasional recurring dreams and flashbacks. But, I never suffered any ill effects from my time there, aside from being spooked by periodic siren testing from a nearby fire station back home. Otherwise, I would just re-live some of the lightly haunting moments that would bubble-up and re-visit my mind every now and then. Of course, my own combat-related work experiences there were much less stressful and not all that traumatic in comparison to the many infantrymen and some others who were quite often engaged in the extreme heat of battle. But, there's just something strangely amazing about these kinds of involvements, at least in my case, whether traumatic or otherwise, when one's own psyche tends to unconsciously hang onto even the unimportant finite details throughout the years. Along with those minor parts to the puzzle, the more significant moments of wartime events were also somehow marked indelibly for me, and filed away to allow for their reappearance at some later date and place in time; even after being 38 years removed.

~ Sixteen ~

Flight of the Freedom Bird

When one's time increasingly drew closer to the point of rotating out of Vietnam, each of us eventually became what was termed, a 'Shortimer', simply because our time from then on out was getting short. For many of us, when that time finally came, thoughts of actually boarding that near mythical 'Freedom Bird' to head back home for good occasionally danced through our heads and entertained us with what seemed to represent the full resumption of our former civilian lives. When the days remaining in country finally whittled down to double digits, the realization that the war would soon be over for some of us became more and more apparent and increasingly more anticipated. From within our field encampments, we would occasionally spot a magnificent 'Freedom Bird' flying high overhead. Pausing momentarily to reflect on it as it passed over, we always clearly noted its distinctive white vapor trails streaming out against the lofty blue beyond. It represented our great hope and salvation; our ticket out of that 'hell hole' to transport us back to reality, or at least to a somewhat less sobering version of it.

In my case, I still had ten months remaining on my 3-year enlistment, before rotating out of 'Nam, but felt that I could pretty much do that time "standing on my head," after eighteen months duty over there. Being an ardently independent-minded person, I had been a somewhat reluctant enlistee, while the military had always seemed a bit like a minimum-security prison to me, with more time left to serve out before finally regaining my long-lost freedom. I didn't complain about it, though; I just kept the low-level resentment of it with me in my inner core and endured the ongoing situation all the while. It was almost like my own little private war within a war, to be eventually won against the Army (*not that it was the Army's fault*). However, at that point, time leaned more on my side, with at least the foreseeable promise of my eventual return to the States then being somewhat more within reach.

Whereas I had my own particular misgivings with the Army through what had long been considered to be my 'predicament' of sorts, others there were fine with the military and its strict control-oriented format. Many of our officers and some of our NCOs even re-enlisted to become what we called "lifers". But, as I had always been somewhat of a maverick, it never really fit with me as any kind of career move. It was simply a coerced situation that I was stuck with for the time being, as I had enlisted under a certain amount of duress. Besides, the pay was so meager that it was really nothing to write home about. Occasionally, some others had also shared that ongoing monetary concern. Sure, the Army took care of our fundamental needs, as we lived and worked at our duty stations, but for the lower ranks, it was always difficult to save much of one's pay, as it didn't amount to a lot each month. However, despite the few frustrations and incidental grumblings that we endured as we spent our time there, most everyone got along well enough, and we tended to keep ourselves focused more on our land clearing missions than on the little annoying distractions that we really had no control over.

As a recognized 'Shortimer', I had been approached by recruiters on a few occasions with proposals on possible re-enlistment for an additional six-year term. Their renewal campaigns there mainly focused on those men with a minimal amount of time remaining on their existing enlistment. Nevertheless, as much as they tried and as much as they offered, their efforts always fell on deaf ears. For me, it didn't matter how much money or rank was dangled in my direction. The answer was always, "No, sir!"

My original 3-year commitment to the Army was really all that I could ever allow myself to endure, as the military was really not a place to fully 'be yourself'. Most decisions were usually made for you through the established chain of command, while any and all orders were not to be questioned in any way, shape or form, without facing some sort of retaliatory punishment for it. In that regard, I suppose that I had always been somewhat of a misfit, seeing as I viewed the Army at the time, as an archaic and somewhat draconian type of military system. But, at least I resolved to see things through, despite my situation and, as much as possible, tried to be a team player, as we worked our lengthy operations in the field. It just wasn't entirely enjoyable and I know that it wasn't supposed to be. I did learn a great deal more about operating and repairing machinery, though, and managed to shake off my earlier naïve ways there in South Vietnam as the stress of war, along with our grueling work, had forcibly pushed me ever closer toward mental maturity. But, in the final analysis, I just wanted my freedom back. I had grown tired of my semi-forced captivity, and had come to value my personal liberty above all else.

Fortunately, my time in Land Clearing wasn't fraught with spit & polish, KP, and guard duty inspections, compared with stateside duty, as I had always privately detested that particular aspect of Army life. Instead, it was full of demanding, hard work, with long hours and plenty of adventurous moments, as well as some potentially perilous ones. For the most part, though, our officers and higher ranking NCOs tended to relax the rules somewhat as they guided our field operations, and mainly left us to accomplish our hazardous, and often difficult, daily routines.

During my last Stand Down, when my time in Vietnam had finally dwindled down to about a month to go, I went over to our company's supply room and rescued the large ceramic vase that I found up in Song Be from our cigar-chomping supply sergeant. He was true to his word in keeping it there for me, unmolested, although the interior had to be cleared of ash and cigar butts. From there, I transported it across Long Binh Post to our 'Whole Baggage Point', where several Vietnamese locals were employed in the task of packaging large items to be sent back to the States. All stateside-bound personnel were given an allowance for this, limited to 180 lbs. of declared Whole Baggage, for special handling of large or heavy items that could not otherwise go on the commercial jetliners. The workers there all looked the vase over in amazement, seemingly knowing that it was indeed a special piece. They reacted vocally in their sing-song-like chatter and packed it up with care in a wooden crate for the long journey out across the wide Pacific pond. It was the one physical item in my possession that I took precautions to guard its safety, as I regarded it as a true souvenir of Vietnam, and a very unique and unusual treasure piece of sorts.

Several years later, it was privately appraised, and determined to be of Chinese origin. Given that there were many Chinese who had earlier migrated to Vietnam, and the fact that the vase was found at this kiln site in Song Be, it was easy to deduce that it was more likely to have been turned and fired there by an artisan of Chinese influence, than to have been imported from somewhere in China. At this writing, it remains on display in my 94-year-old mother's living room, as a decorative artifact that has become a treasured and highly coveted family possession.

Dragon vase measures out at 34 ½ "tall, x 22" wide, w/top diameter at 12 ¾".

I had other accumulated property that was acquired or purchased over there, like a reel-to-reel tape deck, turntable, and other stereo equipment. But, everything had previously been boxed up and shipped out through the mail a few months ahead of time. As it was, the vase was the very last item to go, short of getting my own olive drab tail feathers the hell out of there.

With my time dwindling down to just a week before finally rotating out, I departed our field NDP encampment for the last time and returned to Long Binh. At that time, I remained on duty within the company area, while also taking care of any other loose ends that might have needed further attention. It gave me the time that I needed to sort out and pack up all of my remaining personal essentials, while still working on some light maintenance within the motorpool. I could hardly believe that the moment was suddenly so near, after what had seemed like an eternity of time spent within the war-torn confines of III and IV Corp Vietnam. When that day finally came, I said goodbye to the 501st, and caught a ride out of Long Binh on a relatively short road trip over to Bien Hoa, where I would process out and leave the Country, just as I had originally arrived, 18 months prior.

Upon arrival in Bien Hoa, I checked into the very same holding company that had previously processed me into the Country, and waited there for my new duty station orders to be written up and placed in hand. I stayed there that night, anticipating my 'wake-up' day, and found myself gravitating to the exact same corner bunk in the barracks that I had slept on a year and a half earlier. It just seemed fitting. The next day after breakfast, my name was called out in formation, and I received my written orders from the presiding staff sergeant there. Curious as could be, I opened the large manila envelope to find that I was assigned to Fort Riley, Kansas.

Later, during my 30-day leave, I had voiced my discontent with family members about going to Kansas to serve out the remaining 10 months on my 3-year enlistment. As my parents were long time campaign supporters of our Congressman, Jerome Waldie, they had contacted him, unbenounced to me, and requested a possible change in my duty station. A few days later, one of his secretaries called from Washington, DC, and amazingly, gave me three other choices to pick from. They were, Fort Belvoir, Virginia, Fort Polk, Louisiana, and Fort Benning, Georgia. I chose Benning, and later found that a PI (political influence) marker was placed in my 201 file. With that in view, higher-ranking personnel generally resented anyone who received special treatment, especially the political kind that tends to circumvent normal military channels.

As the day progressed, many of us soon boarded buses, riding only a few miles over to Bien Hoa Air Base, where we would soon catch the majestic flight of that celebrated in-country icon of liberty, the 'Freedom Bird'; otherwise known as a Boeing 707. After a lengthy wait for re-fueling and baggage handling, we boarded the plane from the tarmac and soon taxied down the runway, taking off to the uproarious sound of jubilant cheers and applause from all of the anxious and excited men aboard. It was truly a thrilling moment for nearly everyone on the flight. Joyfully savoring the occasion, we openly celebrated our survival from the war, while fully embracing the exciting reality of finally returning home.

After refueling on Okinawa and spending nearly a full days' time enroute, we touched down at Travis Air Force Base in Fairfield, California, which was good news for me, as it was located within the San Francisco Bay area, fairly close to my own hometown of Alamo. Again, we boarded buses; this time for the 45-minute trip to the Oakland Army Base to process out and go home on leave. For some, it was somewhat like graduation day, as they joyfully anticipated their long-awaited discharge from the Army, while optimistically proclaiming their newfound freedom and hopes for the future. Many onboard still had to book their outbound air and ground transportation after processing out of Oakland and hitting the street. In some cases, these men had to extend their travel even further, as some still had to get across the entire length, or breadth of the country before finally arriving home, while my destination was just a mere half-hour away. Fortunately, the Army provided travel pay to help cover any of their additional inbound expenses.

Just as our buses rolled out through the main gate of Travis, we noticed there were quite a few people lined up along the roadway leading out, and it was becoming evident that they weren't of a friendly nature. They immediately got our attention, with most of us standing up on the bus to get a full outward view of the unfolding demonstration. This group turned out to be Vietnam War protesters, most of which were hippy types from the sixties counter-culture. We were clearly identified as their main target and they fully intended to vent their frustrations and hatred in our direction. As we slowly moved through their gauntlet, nearly every imaginable expletive could be heard spewing out from within their machine-gun-like lips. With their veins protruding and nostrils flared, the ugliness of seething hatred could be seen on each of their distorted faces. They hit the buses with their raised placards, flipped us off, and spit in our direction in an unbelievable showing of absolute contempt. This took everyone on our bus aback, as we were somewhat shocked at this unexpected display before us, even though most U.S. soldiers were already well aware of their element of opposition. But, experiencing this full-force type hatred that they so pointedly and vehemently expressed was far more than any of us had ever expected. Although well organized, this was not simply a peaceful anti-war protest, yet no actual violence had occurred. But, it was scornfully more personal in its attack on us, and left a rather foul taste in the mouths of many aboard the bus as we finally rolled clear of them and headed south, toward Oakland. Apparently, we were perceived as the bad guys then, at least in some people's eyes. What had earlier been a very joyful and engaging mood that we all shared after leaving Bien Hoa, suddenly changed to a more somber one. The bus then remained relatively quiet until we all finally got off at the Oakland Army Base.

It was then made crystal clear that there would be no fanfare for any of us coming home from this war, only a bittersweet re-connection with a moody and disillusioned populace. Unlike those who had returned triumphantly from the bloody wars of our fathers and grandfathers, we were then left to feel like our efforts were unappreciated and seen as less than noble. In addition to this new reality, each man who would return from Vietnam also had his own collection of haunting and/or traumatic war-time memories to bear. These lingering effects tended to follow many of those involved in the war like some dark, inner shadow that casts its pall over one's mind from time to time.

As the Vietnam War wound down in the following years and we finally ended our involvement there, it became quite clear to me that we did not actually lose that war, despite all the untruths that the North Vietnamese and the world press were eager to proclaim against us. In the U.S. government's interest of exercising an exit strategy, the Paris Peace Accord was signed in January of 1973, with all parties participating in the mutual agreement. At that time, the war was over, according to this new treaty, and much of our equipment and arms, along with training, were subsequently given over to the ARVN forces (*South Vietnamese Army*). Our troops re-deployed, incrementally, back to the States, or to other overseas duty stations, as we were then handing our part of the job back over to them, to directly secure themselves and eventually determine their own government's destiny.

With regard to the Land Clearing units, many were already training the ARVN Engineers well in advance of the Paris Accord, and were gradually disbursing equipment to them as their newly formed Vietnamese Land Clearing companies were brought up to strength. The 501st, with all of its men and equipment, was fully de-activated on 8 April 1970, with an ARVN Engineer unit getting some of the equipment. The other units within the 62nd battalion received the rest. It wasn't until later in 1971 (*around September*) that the other two remaining Land Clearing companies (*the 60th and the 984th*) were also de-activated and, likewise, their equipment was disbursed mainly toward ARVN interests. The 62nd Engineer Battalion then returned to the States, where they were re-designated as a Combat/Construction Battalion, at Fort Hood, Texas.

On 29 March 1973, according to a book entitled, 'The Vietnam Experience: The False Peace 1972-74' (*Boston Publishing Co., 1985*), the last of our combat troops departed for the States, leaving over 150 Marine Embassy guards behind, along with some remaining uniformed advisory personnel, who were members of the Defense Attaché Office, which was allowed, according to the terms of the Paris Treaty. At some point, in 1975, it had become more evident that the South Vietnamese could no longer effectively secure themselves well enough from an enemy that refused to honor their previous commitment to the 1973 Paris Accord. In the end, our help was needed in transporting many of their refugees out of the country, along with our few remaining Marines and advisers, with a massive helicopter airlift operation taking place out of Saigon. Afterwards, the media and the North Vietnamese suggested that we had actually been defeated militarily; but we were not. Instead, our efforts there were politically stymied by the growing unrest within our own nation.

It was mainly the 'roadblock' type opposition within our own Congress that increasingly moved our government into a non-committal type of position, which ultimately set the stage for our exit strategy through the Paris Accord. After our 1973 final troop withdrawal, this Congress soon passed sweeping legislation that prohibited any further U.S. military involvement in Vietnam, effectively circumventing the President's power of veto. It abruptly cut off funding for the war and stopped any further military-connected assistance to the South Vietnamese.

After that, the North Vietnamese and Viet Cong forces simply swooped in for the kill, since we were then clearly out of their way. They intentionally broke the signed Paris agreement, knowing all too well that we wouldn't follow suit. Our own Congress had effectively tied our hands, and the South Vietnamese were then left to their own devices to fend for themselves and ultimately accept their own sad and dismal fate, short of finding a way out of the country before the Communist's inevitable sweep was complete.

The only reason I'm touching on all of this is because there are still many people in this country (U.S.) who have been largely misinformed and misled by the predominantly left-wing media, who still believe, to this day, that we were defeated by the Northern communists and the Viet Cong, and that we actually lost the war in Vietnam. But, anyone can look up the facts and see for themselves that what I'm saying here is true, despite the purported historical misrepresentations. However, it just might be the case, considering the political leanings of these media types that they simply prefer to harbor their bias for selfish political reasons, insisting on believing otherwise in spite of the actual truth of the matter. I can fully understand the reason for distorting the truth as it pertains to the North Vietnamese; given their long, agonizing period of struggle and suffering against us, along with the uplifting effect that later served to help re-vitalize a dis-heartened people with the alleged defeat of a world super power having been (fictitiously) achieved. But, the media's irresponsible involvement with this insistent, gross misrepresentation of the facts remains a travesty in the eyes of all those who served, and all who knew otherwise.

So, in actuality, we neither won it nor lost it, and we definitely didn't leave in retreat. Whole units were systematically deactivated or re-deployed to other duty stations in the States or elsewhere. Our military was slowly phased out of the region over a period of time, once the Paris Accord had been signed and put into play. We simply did the honorable thing with regard to the treaty, and withdrew our military support, keeping with the mutually agreed upon pact. Our own

Congress made sure of it. Its signing was not a concession of defeat, as some had later suggested; it was simply an agreement in principle, for a lasting peace between the two governments. In fact, at the risk of sounding a little callous, it was the South Vietnamese who actually lost the war, nearly two years after that 1973 withdrawal; not us. Although we appeared to be the principal player up until the moment when the treaty took affect, we were actually an assisting force that had made tremendous progress in helping to drive the enemy back. But, ultimately, we were not allowed to push on toward victory and finish the job that we originally started several years before.

In most instances there, the South depended greatly on us to help solve their ongoing problem with the North, along with the constant harassments that came from the relentless Viet Cong guerillas, who were ruthless southern Communist sympathizers that had represented the first line of attack against our allied forces, with their dogged and determined, covert-like, venomous strikes of insurgency. However, with the eventual exodus of our established forces, the South then found that it was sadly becoming more and more futile for them to save their government and their somewhat democratic way of life. In the end, they were unable to stem the rising tide of opposing forces, and finally succumbed to their dreaded fate. We could only watch from afar and help transport some of their refugees from the desperation and chaos that ensued in Saigon, as the invading communist troops approached the outskirts of the city.

With some 58,000 American service men and women counted among the dead, along with the hundreds of thousands of Viet Cong and North Vietnamese Army losses, South Vietnamese Army losses, and countless civilians, not to mention the maimed and disabled, the Vietnam War proved to be a very costly involvement for all sides. In that regard, the memory of it will long endure, even as things progressively change and are ultimately reshaped through the natural healing hands of time.

In spite of the way in which the war had changed direction, and ended in disappointing fashion, it firmly remains something in which I will always be proud of my own involvement and contribution there, as well as the individual and combined contributions of others. My somewhat lengthy experiences there in the remarkably lush, jungle-covered environment while working with the Rome Plow tractors, were hauntingly memorable and, at the time, were a grueling test of my own stamina and ability, which had proved to serve as a learning vehicle for years yet to come. Now, after these many years of reflection, I can easily say that Land Clearing in Vietnam, with all of its various elements of adversity, discovery, and camaraderie, was truly the greatest high adventure experience that I had ever been involved with. Knowing that so many lives had been saved due to our combined efforts there, it was found to be, in itself, a most gratifying and noble cause, which still mystifies, haunts, and humbles me from time to time, as I continue my inward travels out along the winding open road of life.

As a dedicated group of determined young men, we endured the almost daily discomforts, hardships, and hazards associated with the arduous work of clearing vast areas of jungle-covered terrain, simply by making the best of it along the way. Through it all, we were subtly spurred on by our highly revered engineer motto, 'Essayons', which is French for, 'Let Us Try'.

Epilogue

Many years after the war, with the advancement of computer technology evolving into the advent of the internet, suddenly a myriad of information had been conveniently placed at one's fingertips. As a result, military reunion associations were more readily formed, and commonly established. Many of their otherwise lost members were then located with far less difficulty than in previous years. This applied to us as well, while certain members established an association for our own unit, the 86th/501st Land Clearing Team, along with an all inclusive one, which grouped together all of the Land Clearing Battalions, Teams, Platoons, etc., into the *Vietnam Land Clearer's Reunion Association*.

In the summer of 2000, I jumped in my truck and made the long-awaited trek, traveling nearly halfway across the country to gather with many of my fellow land clearers at the re-designated 62nd Engineer Battalion's home base of Fort Hood, Texas. At that time, we renewed our associations and friendships with one another in our company, while alternately meeting some of the other men from the other units. We then gathered to mark the millennium by dedicating a fully-dressed Rome Plow to serve as a proud reminder of the battalion's unusual, but effective designation during the Vietnam War. The newly converted D7-K Rome

Plow stood like a giant sentry at the front entrance to their battalion area, showcasing a quite remarkable machine that helped to save so many lives during an all but forgotten war, within a nearly bygone era.

Following the dedication ceremony, we were all allowed to board the newer-styled plow and take pictures, etc. This Caterpillar D7-K turned out to be an upgraded version, although quite similar to the ones that we had operated, with the cab being just a bit larger, overall, in size.

6 July 2000 – Fort Hood, Texas

After entering the cab and taking up my old position in the operator's seat, I immediately felt the old familiarity from my earlier days in the jungle. While sitting there taking it all in again, some old familiar memories naturally came drifting back into view.

Suddenly this was the highlight of my trip, as I thoroughly enjoyed the moment during that brief reunion with this upgraded version of the machine that I had earlier come to know so well. After all these many years, I had then finally come back around to nearly a full circle, where this honored occasion seemed entirely appropriate and altogether fitting.

Made in the USA
Middletown, DE
18 October 2022